DIELECTRIC PHENOMENA

IN

HIGH VOLTAGE ENGINEERING

Rough Draft Printing

RDP

DIELECTRIC PHENOMENA

IN

HIGH VOLTAGE ENGINEERING

BY

F. W. PEEK, Jr.

First Edition

Rough Draft Printing

RDP

PREFACE

It is the object of the author to give in this book the properties of gaseous, liquid and solid insulations, and methods of utilizing these properties to the best advantage in the problems of high-voltage engineering. Such problems require a knowledge, not only of the laws and mechanism of breakdown of dielectrics as determined by experiment, but also a simple working knowledge of the dielectric circuit.

Methods that have proved useful in designing apparatus, transmission lines, insulators, bushings, etc., are discussed and illustrated by practical application. In addition, such subjects as the manner of making extensive engineering investigations and of reducing data, the measurement of high voltages, the effects of impulse and high-frequency voltages, methods of drawing dielectric fields, outline of modern theory, various dielectric phenomena, etc., are considered. In all cases where laws and discussions of dielectric phenomena are given, it has been thought best to accompany these with experimental data.

Much original work is given, as well as reference to other investigations. The author's extensive research was made possible by facilities afforded by the Consulting Engineering Department of the General Electric Company, for which acknowledgment is made. Thanks are due Mr. H. K. Humphrey, and others, who have greatly assisted in the experiments and calculations.

F. W. P., Jr.

Schenectady, N. Y.,
April, 1915.

RDP

CONTENTS

CHAPTER IX

CHAPTER X

CHAPTER XI

ISBN 1-60386-112-2

DIELECTRIC UNITS

Electromotive force, volts	e volts.
Gradient	$g = \frac{e}{x}$ volts/cm.
Permittance or capacitance or capacity.	$C = \frac{kKA}{x} = 8.84 \frac{kA}{x} 10^{-14}$ farads.
Permittivity or specific capacity	
relative	k ($k = 1$ for air)
absolute (air)	$K = \frac{10^9}{4\pi v^2} = 8.84 \times 10^{-14}$ farad cm. cube.
Elastance	$S = \frac{1}{C}$
Elastivity	$\sigma = 1/k$
Flux, displacement	$\Psi = Ce = \frac{e}{S}$ coulombs (or lines of force).
Flux density,	$D = kKg$ flux or displacement per cm.^{2}
Intensity	F (unit not used in text).
Stored energy	$w_e = \frac{Ce^2}{2}$ joules.
Energy density	$w_o = \frac{gD}{2}$ joules per cm. cube.
Permittance or capacity current	$i_c = \frac{d\psi}{dt} = C\frac{de}{dt}$ amps.
Permittance or capacity current	$i_c = 2\pi fCe$ amps. for sine wave.
Permittance in series	$\frac{1}{C} = \frac{1}{C_1} + \frac{1}{C_2} + \frac{1}{C_3}$
Elastance in series	$S = S_1 + S_2 + S_3$
Permittance in multiple	$C = C_1 + C_2 + C_3$
Elastance in multiple	$\frac{1}{S} = \frac{1}{S_1} + \frac{1}{S_2} + \frac{1}{S_3}$

v = velocity of light $= 3 \times 10^{10}$ cm. per sec.
x = spacing cm. A = area in sq. cm.

NOTE.—For non-uniform fields e, x, etc. are measured over very small distances and become de, dx, etc. Then the gradient at any point is $g = \frac{de}{dx}$, etc.

TABLE OF SYMBOLS

The following is a list of the principal symbols used. The use given first is the most general one. The meaning is always given in the text for each individual case.

A area in square cm., constant.
$A'_1A'_2$ flux foci or flux centers.
a distance, constant.
b barometric pressure in cm., constant, distance.
C permittance or capacity.
$C_{r_1r_2}$ permittance between points r_1 r_2.
C_n permittance to neutral.
c constant, distance.
D dielectric flux density.
d distance, constant.
e voltage.
e_n voltage to neutral.
$e_{r_1r_2}$ voltage between points r_1 r_2.
e_p voltage to point p.
e_v visual critical corona voltage.
e_o disruptive critical corona voltage.
e_d disruptive critical corona voltage for small wires.
e_s spark-over voltage.
f frequency.
f, f_1, f_o coefficients used in reducing average gradient to maximum—see page 28.
F constant (sometimes used for dielectric field intensity).
g, G gradient.
g gradient volts per cm. or kilovolts per cm.
g gradient volts per mm. for solid insulations.
g_v visual critical gradient.
g_o disruptive critical gradient.
g_d disruptive critical gradient for small wires.
g_{max} maximum gradient—see note below.
g_s spark gradient.
g_a gradient at point a.
h constant, height.
i current amperes.
K dielectric constant for air

$$K = \frac{10^9}{4\pi v^2} = 8.84 \times 10^{-14} \text{ farads per cm. cube.}$$

k relative permittivity ($k = 1$ for air).
L inductance.
l length, thickness

M constant.
m ordinate of center of line of force.
m mass.
m_v, m_o irregularity factor of conductor surface.
N neutral plane.
n number.
O center point.
P point.
p power loss.
q constant.
r radius of wires or cables.
R radius of spheres, of outer cylinder.
r resistance.
S, s spacing between conductor centers.
S' distance between flux foci.
S elastance—see page 11.
t temperature, thickness.
T time.
v velocity of light in cm./sec. $= 3 \times 10^{10}$.
v velocity.
w_l magnetic stored energy.
w_c dielectric stored energy.
w weight.
X, x cm. spacing between conductor surfaces, thickness, coordinate of a point.
x_1, x_2 distance.
y coordinate of point.
z distance from the center of a conductor or an equipotential circle to flux foci.
α angle, constant.
β constant.
δ relative air density.
$\Delta\Sigma$ difference of two sums.
ϵ base of natural log.
Ψ dielectric displacement or dielectric flux.
Φ magnetic flux.
ϕ angle, function.
θ angle.
σ elastivity.
Σ sum.
$\Sigma\Sigma$ sum of two sums.
ω resistance.
mm. millimeter.
cm. centimeter.
$\cong$ approximately equal to.

Note that voltages in measured data are often given to neutral; in such cases the single phase line to line voltages are twice (2), and the three phase (symmetrical) $\sqrt{3}$ times, these values.

Permittances or capacities are also frequently given to neutral because it is a great convenience in making calculations.

The subscript max. is often used to distinguish between the maximum and root mean square or effective. This is done because insulation breakdown generally depends upon the maximum point of the wave. Such voltages may be reduced to effective sine wave by dividing by $\sqrt{2}$. Sometimes when the maximum gradient is referred to it means the gradient at the point in the field where the stress is a maximum. These references are made clear in the text for each individual case.

Tests were made on single-phase lines unless otherwise noted.

Views or theories advanced by the author are always accompanied by sufficient experimental data so that the reader may form conclusions independently.

RDP

DIELECTRIC PHENOMENA

CHAPTER I

INTRODUCTION

It is our work as engineers to devise means of transmitting energy electrically, from one point to another point, and of controlling, distributing, and utilizing this energy as useful work. Conductors and insulating materials are necessary. Transmission problems are principally problems of high voltage and therefore of dielectrics. In order that energy may flow along a conductor, energy must be stored in the space surrounding the conductor. This energy is stored in two forms, electromagnetic and electrostatic. The electromagnetic energy is evinced by the action of the resulting stresses, for instance, the repulsion between two parallel wires carrying current, the attraction of a suspended piece of iron when brought near the wires, or better yet, if the wires are brought up through a plane of insulating material, and this plane is dusted with iron filings, and gently tapped, the filings will tend to form in eccentric circles about the conductors. These circles picture the magnetic lines of force or magnetic field in both magnitude and direction. This field only exists when current is flowing in the conductors. If now potential is applied between the conductors, but with the far ends open circuited, energy is stored electrostatically. The resulting forces in the dielectric are evinced by an attraction between the conductors; a suspended piece of dielectric in the neighborhood is attracted. If the conductors are brought through an insulating plane as before, and this is dusted with a powdered dielectric, as mica dust, the dust will tend to form in arcs of circles beginning on one conductor and ending on the other conductor. See Fig. 1(*a*) and (*b*). The dielectric field is thus made as tangible as the magnetic field. Fig 1(*c*) is an experimental plot of the magnetic and dielectric fields. Fig. 1(*d*) is the mathematical plot. Fig. 1(*c*) represents the magnetic and dielectric fields in the space surrounding two conductors which are carrying energy. The power is a function of the product of these two fields and the angle between them. In comparing Figs. 1(*c*) and (*d*) only the general direction and relative density of the fields at different points can be considered. The actual number of lines in Fig. 1(*c*) has no definite meaning. The dielectric lines of force in Fig. 1(*d*) are drawn so that one twenty-fourth of the total flux is included between any two adjacent lines. Due to the dielectric fields, points in space surrounding the conductors have definite potentials. If points of a given potential are connected together, a cylindrical surface is formed about the conductor; this surface is called an equipotential surface. Thus, in Fig. 1(*d*), the circles represent equipotential surfaces. As a matter of fact, the intersection of an equipotential surface by a plane at right angles to a conductor coincides with a magnetic line of force. The circles in Fig. 1(*d*), then, are the plot of the equipotential surfaces and also of the magnetic lines of force. The equipotential surfaces are drawn so that one-twentieth of the voltage is between any two surfaces. For example: If 10,000 volts are placed between the two conductors, one conductor is at +5000 volts, the other at −5000 volts. The circle (∞ radius) midway between is at 0. The potentials in space on the different equipotential surfaces, starting at the positive surfaces, are +5000, +4500, +4000, +3500, +3000, +2500, +2000, +1500, +1000, +500, 0, −500, −1000, −1500, −2000, −2500, −3000, −3500, −4000, −4500, −5000. A very thin insulated metal cylinder may be placed around an equipotential surface without disturbing the field. If this conducting sheet is connected to a source of potential equal to the potential of the surface which it surrounds, the field is still undisturbed. The original conductor may now be removed without disturbing the outer field.

The dielectric lines of force and the equipotential surfaces are at right angles at the points of intersection. The dielectric lines always leave the conductor surfaces at right angles. The equipotential circles have their centers on the line passing through the conductor centers, the dielectric force circles have their centers on the neutral line.

Energy does not flow unless these two fields exist together—for instance, if the dielectric field exists alone it is aptly spoken of as "static."

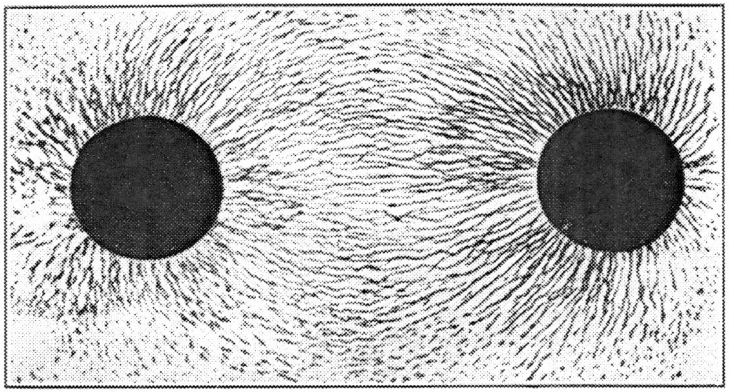

FIG. 1(a).—A photograph of a mica-filing map of the dielectric lines of force between two cylinders.

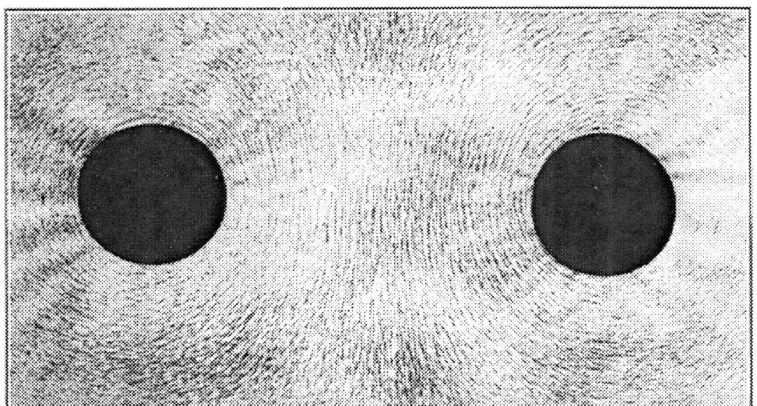

FIG. 1(b).—A photograph of an iron-filing map of the magnetic lines of force about two cylinders.

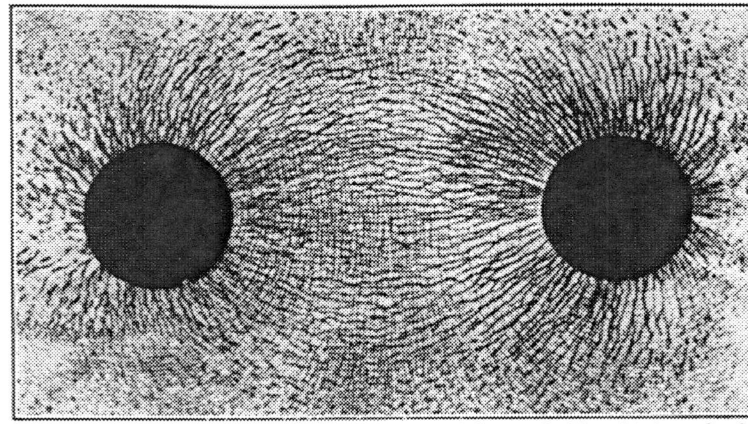

FIG. 1(c).—A photographic superposition of Fig. 1(a) and (b) representing the magnetic and dielectric fields in the space surrounding two conductors which are carrying energy.

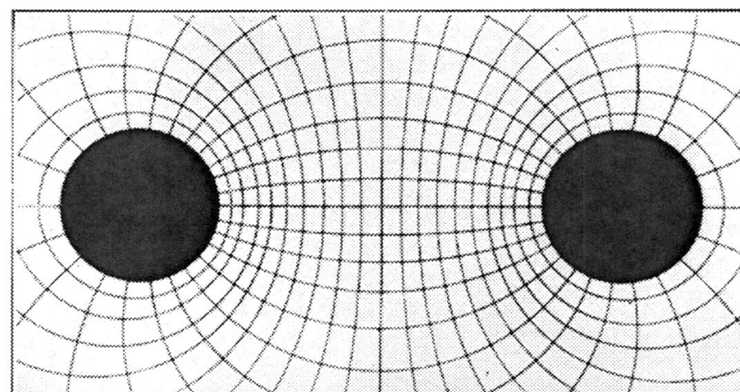

FIG. 1(d).—A mathematical plot of fields shown in Fig. 1(c).

(*Facing page 2.*)

The energy stored in the dielectric field is

$$\frac{e^2C}{2}$$

where e is the voltage and C a constant of the circuit called the permittance (capacity) and the energy stored in the magnetic

$$\frac{i^2L}{2}$$

field is where i is the current and L is a constant of the circuit called the inductance.

The energy stored in the dielectric circuit is thus greater for high voltage, and in the magnetic circuit for high currents.

When energy was first transmitted, low voltages and high currents were used. The magnetic circuit and magnetic field in this way became known to engineers, but as little trouble was had with insulation, the dielectric field was therefore not generally considered. If insulation broke down, its thickness was increased without regard to the dielectric circuit.

A magnetic circuit is not built in which the magnetic lines are overcrowded in one place and undercrowded in another place—in other words, badly out of balance. Since voltages have become high it is of great importance to properly proportion the dielectric circuit. Although an unbalanced magnetic field may mean energy loss, an unbalanced or too highly saturated dielectric field will mean broken down insulation.

The dielectric and magnetic fields may be treated in a very similar way.[1] For instance, to establish a magnetic field a magneto-motive force is necessary; to establish a dielectric field an electro-motive force is necessary. If in a magnetic circuit the same flux passes through varying cross sections, the magneto-motive force will not divide up equally between equal lengths of the circuit. Where the lines are crowded together the magneto-motive force per unit length of magnetic circuit will be larger than where the lines are not crowded together. The magneto-motive force per unit length of magnetic circuit is called magnetizing force. Likewise for the dielectric circuit where the dielectric flux density is high a greater part of the electro-motive force per unit length of circuit is required than at parts where the flux density is low. Electro-motive force or voltage per unit length

[1] See Karapetoff, "The Magnetic Circuit," and "The Electric Circuit." Steinmetz, "Electric Discharges, Waves and Impulses."

of dielectric circuit is called electrifying force, or voltage gradient. If iron or material of high permeability is placed in a magnetic circuit the flux is increased for a given magneto-motive force. If there is an air gap in the circuit the magnetizing force is much greater in the air than in the iron. If a material of high specific capacity or permittivity, as glass, is placed in the dielectric circuit, the dielectric flux is increased. If there is a gap of low permittivity, as air, in the circuit, the gradient is much greater in the air than in the glass. The electric circuit is also analogous, as will appear later.

A given insulation breaks down at any point when the dielectric flux density at that point exceeds a given value. It is thus important to have uniform density. The flux, ψ, depends upon the voltage, the permittivity, or specific capacity of the insulation, and the spacing and shape of the terminal. That is,

$$\psi = Ce$$

The flux density D, at any point, is proportional to the gradient g, or volts per centimeter at that point, and to the permittivity of the dielectric. Thus,

$$D = \frac{de}{dx} Kk = gKk$$

also

$$D = \frac{\psi}{A}$$

As the density is proportional to the gradient, insulations will, therefore, also rupture when the gradient exceeds a given value; hence if the gradient is measured at the point of rupture it is a measure of the strength of the insulation. The strength of insulation is generally expressed in terms of the gradient rather than flux density.

By analogy with Hooke's Law the gradient may be thought of as a force or stress, and the flux density as a resulting electrical strain or displacement. Permittivity, k, then, is a measure of the electrical elasticity of the material. Energy is stored in the dielectric with increasing force or voltage and given back with decreasing voltage. Rupture occurs when the unit force or gradient exceeds the elastic limit. Of course, this must not be thought of as a mechanical displacement. In fact, the actual mechanism of displacement is not known.

When two insulators of different permittivities are placed in series with the same flux passing through them, the one with the lower permittivity or less electrical elasticity must take up most of the voltage, that is, the "elastic" one may be thought of as "stretching" electrically and putting the stress on the electrically stiff one. The dielectric circuit is also analogous to the electric circuit when the flux is thought of in place of the current—and the permittance as conductance. The reciprocal of the permittance is sometimes called the elastance (S) and corresponds to "resistance" to the dielectric flux. It is convenient when permittances are connected in series, as the total elastance is the sum of the elastances. When permittances are connected in multiple, the total permittance is the direct sum of the permittances. Take two metal plates in air and apply potential between them until the flux density is almost sufficient to cause rupture. Now place a thick sheet of glass between the plates; the permittance and therefore the total flux is increased. This increases the stress on the air, which breaks down or glows. The glass does not break down. Thus by the addition of insulation the air has actually been broken down. This takes place daily in practice in bushing, etc., and the glow is called static.

It is especially important in designing leads and insulators immersed in air to avoid overstress on the air.

It can be seen that a statement of volts and thickness does not determine the stress on the insulation. The stress on insulation does not depend altogether upon the voltage, but also upon the shape of the electrodes; as, for instance, for needle points the flux density at the point must be very great at fairly low voltages, while for large spheres a very high voltage is required to produce high flux density. For this reason 200 kv. will strike 55 cm. between needle points, while it will strike only about 17 cm. between 12.5-cm. spheres. From the above it can be seen that it is much more important to design the dielectric circuit for proper flux distribution than the magnetic circuit. Local overflux density in the magnetic circuit may cause losses, but local overflux density in the dielectric circuit may cause rupture of the insulation.

Consider now the two conductors of a transmission line with voltage between them. The total dielectric flux begins on one conductor and ends on the other conductor (see Fig. 1(d)). The flux is dense at the conductor surface and less so at a distance from the conductor. Hence the voltage gradient is greatest at the surface, where the dielectric cross-section is a minimum, and therefore the "flux resistance" or elastance is greatest and breakdown must first occur there. For the particular case shown in Fig. 2 one-third of the voltage is taken up by the space 12 cm. from each conductor, although the total space is 100 cm. The gradient is greatest at the wire surface. That is, if across a small distance, X_1, the voltage is measured near the wire surface and then again across the same space X_2 some distance from the wire, it is found that the voltage is much higher across the small space X_1 near the wire surface than across the one farther out. In actually measuring the gradient, or rather calculating it, X is taken very small or dx. The voltage across dx is de. The purely mathematical expression for the gradient at the surface of parallel wires is:

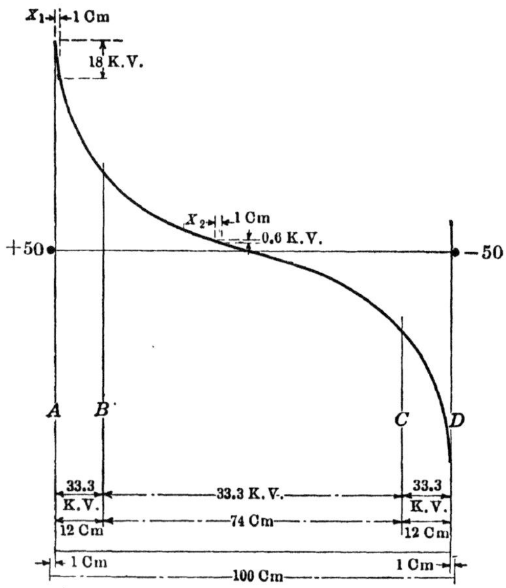

FIG. 2.—Voltage in space between two parallel wires.

$$g = \frac{de}{dx} = \frac{e}{2r \log_e \frac{s}{r}}$$

If the conductors are close together a spark jumps across when the voltage is high enough to produce overflux density at the conductor surface; or corona and spark-over are simultaneous. If far apart, corona forms around the conductor surface and sparkover takes place at some higher voltage.

As voltages or electro-motive forces become higher the proper shaping and spacing of the conductors to prevent dielectric flux concentration becomes of more importance. The dielectric field must now be considered in the design of apparatus as the magnetic field has been considered. Certain phenomena always exist which go unnoticed because of their feeble effect, but which when conditions are changed, usually in a way to cause a greater energy density, become the controlling features. This is so with the dielectric circuit. The problem first made itself apparent to engineers in the transmission line, which will be taken to illustrate this. When voltages were below about 60,000 the conductors used had sufficient radius or circumference so that the surface flux density or gradient was not sufficient to cause breakdown. As voltages became higher the sizes of conductors remained about the same, and therefore the flux density or gradient became greater. The air broke down and caused the so-called corona and resulting loss.

As high voltage engineering problems will be, to a great extent, problems of the dielectric circuit, this will be discussed in the next chapter and calculations made for a few common forms of electrodes. The determination of the dielectric flux density, etc., is purely a mathematical problem. Exact calculations are difficult and often impossible except for simple forms. Exact calculations are not necessary in practical design work, but the general principles must be kept in mind.

CHAPTER II

THE DIELECTRIC FIELD AND DIELECTRIC CIRCUIT

If two conductors placed in a dielectric, as, for instance, the two parallel wires of a transmission line in air, are connected together at one end by an electric motor, or resistance and potential is applied across the other end from an a.c. generator or other source of power supply (see Fig. 3), energy transfers take place. The motor at the far end turns and part of the energy is thus used as useful work—part appears as heat in the motor.

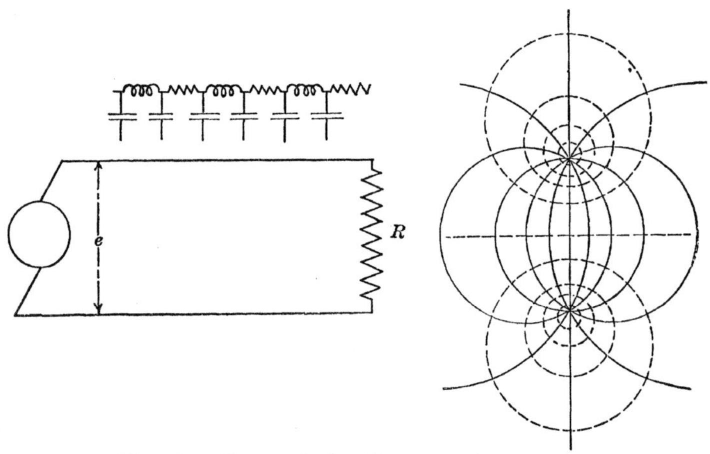

FIG. 3.—Transmission line carrying energy.

As a function of the current in the transmission circuit and a constant called the resistance, energy is absorbed. This energy appears as heat in the conductors; it is proportional to the product of the square of the current and the resistance, and is commonly known as the I^2r loss. Hence, as it is not returned to the circuit or transferred into useful work, but is dissipated as heat, it is *analogous* to a *friction loss.* During the transmission, energy is stored in the space surrounding the conductors in the electric field in two different forms—magnetic and dielectric.

Energy is stored in the magnetic field, where it is proportional to the square of the current and to a constant of the circuit called the inductance:

$$w_1 = \frac{I^2L}{2}$$

Magnetic energy is stored with increasing current and delivered back to the circuit with decreasing current. The magnetic energy becomes noticeable or large when the currents are large, or in low voltage circuits.

Due to the dielectric field, the energy is

$$w_c = \frac{e^2C}{2}$$

This energy is stored with increasing voltage and delivered back with decreasing voltage. A dielectric may thus by analogy be thought of as an *electrically elastic material*, which is displaced by an electric pressure, *i.e.*, voltage. Energy is hence stored in the dielectric with increasing voltage or electric pressure, is maximum at the maximum point of the voltage wave and is delivered back to the circuit with decreasing voltage. When the pressure becomes too great the *electric "elastic limit"* is exceeded, or the dielectric becomes distorted beyond this "elastic limit," and rupture occurs. The dielectric energy becomes of great importance at high voltage, and hence in the study of insulations, and it only will be considered here. The electric displacement may be pictured in magnitude and direction by lines of force. The dielectric lines of force for two parallel conductors are shown in Fig. 3, the eccentric circles (dotted) are the magnetic lines of force. The magnetic circles are also equipotential boundary lines for the dielectric field. The dielectric energy is sometimes said to be due to a charge on the conductor. This is often confusing, as the energy is stored not

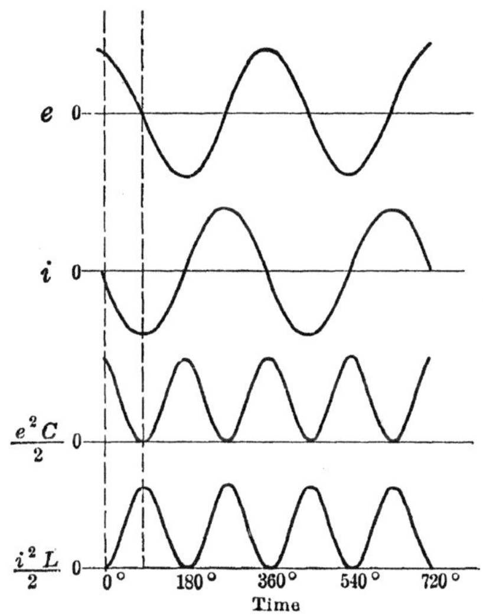

FIG. 4.—Variation of dielectric and magnetic stored energy with voltage and current.

on the conductor but in the surrounding space, and may be thought of as due to an electric displacement. The nature of this displacement is not known. It can be seen that in order that a transfer of energy may take place, energy must be stored in the space surrounding the conductors in two forms—magnetic and dielectric. Energy thus flows only in a space in which there is a magnetic and a dielectric field. This energy is proportional to the product of the magnetic and dielectric field intensity and the sine of the included angle. If one field exists alone there can be no energy flow. The change of stored energy from magnetic to dielectric and back is shown in Fig. 4.

Dielectric Field between Parallel Planes.—In the dielectric field the flux or total displacement is

$$\psi = Ce \text{ coulombs (or lines of force)} \tag{1}$$

where e is the applied e.m.f. or voltage, and C is a constant of the circuit, depending upon its dimensions, and is called the capacity,

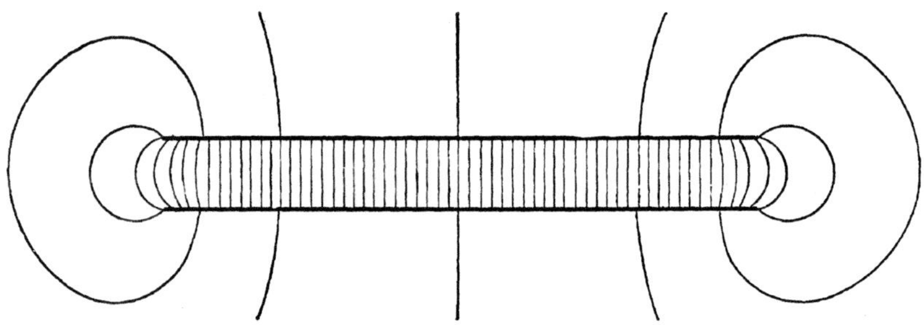

FIG. 5.—Dielectric field between parallel planes.

or better, permittance. If C is measured in farads, and e in volts, ψ is expressed in coulombs. Fig. 5 shows the simplest form of dielectric circuit. Neglecting the extra displacement at the edges, it is seen that the dielectric lines of force are everywhere parallel and the field is uniform. The dielectric circuit constant, or the permittance, is directly proportional to the area of the cross-section perpendicular to the lines of force, inversely proportional to the spacing along the lines of force, and directly proportional to the dielectric constant or the permittivity.

For large parallel planes without flux concentration at the edges

$$C = \frac{A}{X} k \left(\frac{10^9}{4\pi v^2}\right) = \frac{A}{X} k K \text{ farads} \tag{2}$$

where A is the area in square centimeters,
X is the distance between plates in centimeters,
k is the specific inductive capacity, or better, permittivity, and
v is the volocity of light, 3×10^{10} cm., per second.

The term in brackets is due to units. The flux density then, or displacement per unit area, as the flux is uniform, is:

$$D = \frac{\psi}{A} = \frac{Ce}{A} = \frac{ek10^9}{x4\pi v^2} \text{ coulombs per cm.}^2 \qquad (3)$$

To establish this flux or displacement through the distance X an electromotive force e is required. The force per unit length of dielectric circuit or *electrifying force* is constant in the uniform field and is then $\frac{e}{X}$. The gradient then is

$$g = \frac{e}{X} \text{ volts/cm.}$$

The density may thus be written:

$$D = kg\frac{(10^9)}{4\pi v^2} \qquad (4)$$

which is analogous to Hooke's law in Mechanics,

strain = k times stress.

The larger k is, the greater the displacement is for a given force g.

Thus k is the coefficient indicating *electrical elasticity* of the material, or its "conductivity" to the flux. The reciprocal of permittivity is analogous to resistivity and has been termed elastivity (σ). The reciprocal of permittance has been termed elastance (S). The dielectric circuit then becomes analagous to the electric circuit

$$\text{Flux} = \frac{\text{volts}}{\text{“flux resistance”}} \text{ or } \psi = \frac{e}{S}$$

It is often convenient to consider the dielectric circuit in this way, and to use σ and S, as the total elastance of a number in series is the direct sum. The total permittance of a number in multiple is the direct sum. See two methods, Case 1, Chapter X.

In studying insulations it is important to be able to express their relative strengths. This is naturally generally done in terms of the force or voltage gradient necessary to cause rupture. It may also be done in terms of the flux density at rupture; that is, in coulombs per square centimeter. A given insulation breaks down at any point when the flux density exceeds a certain definite value at that point, or when the gradient exceeds a given definite value.

For Fig. 5 where the field is uniform the gradient is

$$g = e/X \text{ kv. per cm.} \quad D = \frac{\psi}{A} = \frac{Ce}{A} = K \text{ kg.} \quad \text{coulombs per cm}^2.$$

$$K = \frac{10^9}{4\pi v^2} = 8.84 \times 10^{-14}$$

Hence, if the voltage is increased until rupture occurs, and found to be e, the voltage gradient or the flux density at rupture is known; it would seem that this would be a good form of test piece with which to study insulation. This is not usually the case because of the extra displacement at the edges which is difficult to calculate. This may, however, be made very small at small spacings by proper rounding of the edges. Equations for the voltage gradient, permittance, etc., will now be given for a few of the common electrodes. In general, calculations are made in the same way, except that the field is usually uniform over only very small distances. The total capacity is found by taking the capacities over distances so small that conditions are still uniform and integrating.

Concentric Cylinders.—Concentric cylinders make a convenient arrangement for studying dielectric strength, especially that of air and oil. On account of the symmetrical arrangement, the dielectric circuit is readily calculated. For testing, the extra displacement at the ends is eliminated by belling (see Fig. 6).

Permittance or Capacity.—In this case the lines of force are radial. The equipotential surfaces are concentric cylinders. The total flux per centimeter length of cylinder is

$$\psi = Ce$$

The permittance may be thought of as made up of a number of permittances in series between r and R, each permittance being between two equipotential surfaces dx centimeters apart.

For a number of permittances in series

$$\frac{1}{C} = \frac{1}{C_1} + \frac{1}{C_2} + \cdots$$

FIG. 6.—Concentric cylinders.

The permittance of the condenser of thickness dx, over which the field is uniform (Fig. 6), per centimeter length of cylinder is

$$dC = \frac{A}{dx}\frac{k(10^9)}{4\pi v^2} = \frac{2\pi x}{dx}\frac{k10^9}{4\pi v^2} = \frac{k10^9}{2v^2}\frac{x}{dx}$$

$$\frac{1}{C} = \int_{x=r}^{x=R}\frac{1}{dC} = \frac{2v^2}{k10^9}\int_{x=r}^{x=R}\frac{dx}{x} = \frac{2v^2}{k10^9}\log_e R/r$$

$$C = \frac{k10^9}{2v^2 \log_e R/r} = \frac{5.55\ k10^{-13}}{\log_e R/r} \text{ farads per centimeter length of cylinder.} \tag{5}$$

Gradient and Flux Density.—The flux density is greatest at the conductor surface and, hence, the gradient must be greatest there. The flux density at any point x measured from the center is

$$D = \frac{\psi}{A} = \frac{Ce}{A} = \frac{ek}{A2v^2 \log_e R/r}10^9$$

$$A = 2\pi x$$

$$\therefore D = \frac{ek}{4\pi x v^2 \log_e R/r}10^9 = 0.884\frac{ek}{x \log_e R/r}10^{-13} \text{ coulombs per cm.}^2$$

$$\therefore g = D\frac{4\pi v^2}{k}10^{-9} = \frac{e}{x \log_e R/r} \text{ kv. per cm.} \tag{6}$$

g is maximum at the surface of the inner cylinder or where $x = r$

$$g_{max} = \frac{e}{r \log_e R/r} \text{ kv./cm.} \tag{6a}$$

Parallel Wires.—Parallel wires, one of the most common practical cases, will be considered in detail, in order to illustrate the general method of calculating the dielectric circuit and to show that the expressions for the permittance, flux density, gradient, etc., are quite simple and can be written with the aid of ordinary geometry and calculus.

Equipotential Surfaces, Lines of Force, and Flux Density.—All of the equipotential surfaces which arise in this case are cylindrical. Therefore, only their intersections with a normal plane need be considered, and the problem may be dealt with as affecting only the plane.

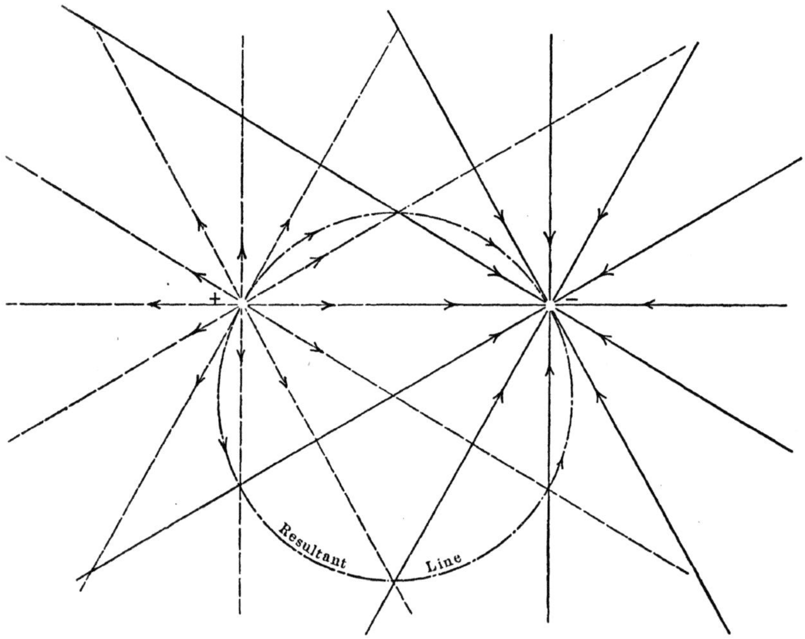

FIG. 7.—Lines of force between parallel wires, by superposition of independent radial fields.

The following principles will be used:

(1) The resultant field in the space between two conductors is the superposition of the two independent fields. The resultant field due to any number of fields may be found by combining in pairs.

Fluxes may be added directly.

(2) The potential at any point is the *algebraic* sum of the potentials due to the independent fields through that point. In the same way, the potential difference between two points

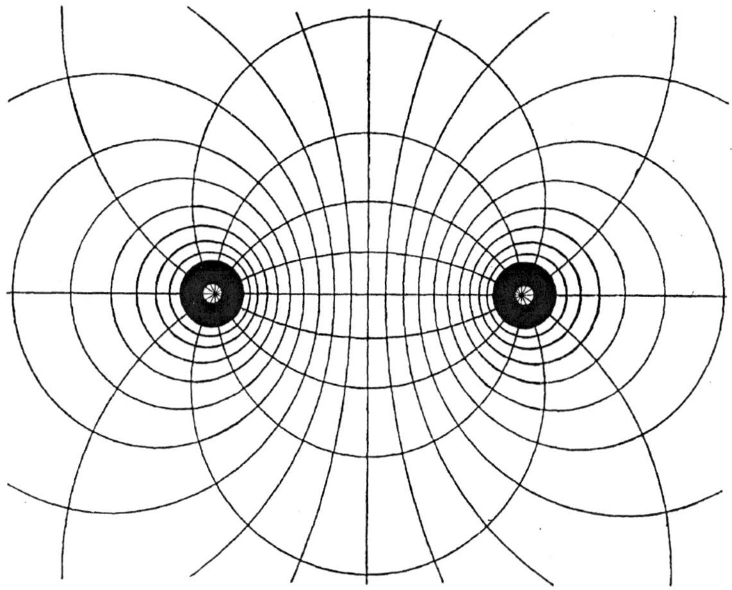

FIG. 8(a).—Lines of force and equipotential surfaces between parallel wires.

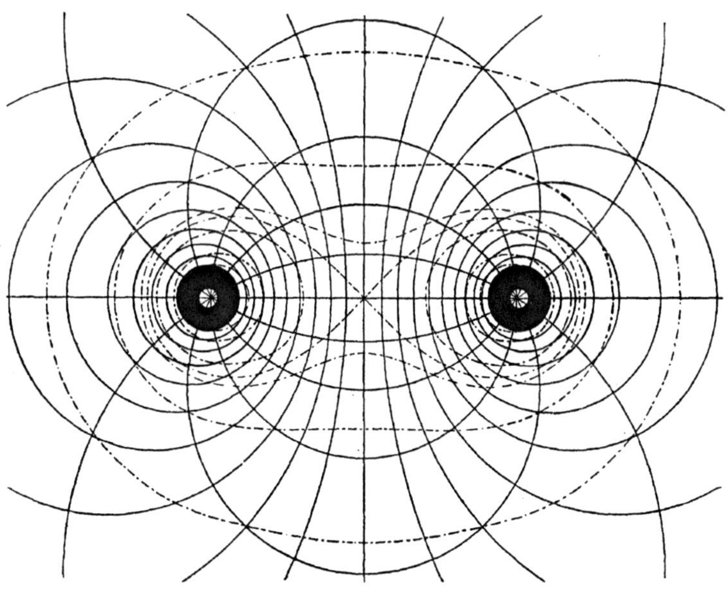

FIG. 8(b).—Lines of force, equipotential surfaces, and equigradient surfaces between parallel wires.

is the *algebraic* sum of the potential differences due to the independent fields.

(3) The density or gradient at a point is the *vector sum* of the densities or gradients due to the independent fields.

When the conductors are infinitely small, the dielectric field may be considered as that resulting from the superposition of the two uniform radial[1] fields from the conductors to an infinite cylinder.

The resultant equipotential surfaces are then cylinders whose right sections are eccentric circles which enclose the wires, and whose centers all lie in the line connecting them; and the lines of force are arcs of circles intersecting in the conductors. The independent radial fields about each conductor are shown in Fig. 7, and the field resulting by superposition is shown in Fig. 8.

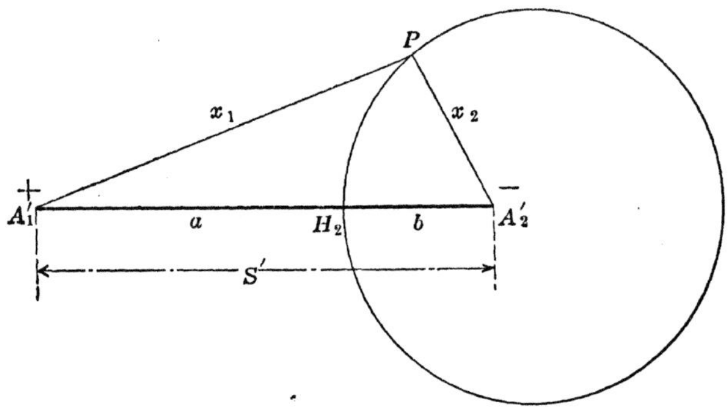

FIG. 9.

The equipotential surfaces will first be considered. It has been shown above that the permittance C between two equipotential cylinders of radii R and r is

$$C = \frac{2\pi kK}{\log_e R/r} \text{ farads per cm.} \qquad (5)$$

This is also the permittance between any two points on these surfaces; therefore, the voltage between points distant r_1 and r_2 cm. from the conductor center is

$$e_{r_1r_2} = \psi/C_{r_1r_2} = \frac{\psi \log_e r_1/r_2}{2\pi kK}$$

Considering now the field resulting from superposition: In

[1] By uniform radial field is meant one in which equal central angles always include equal fluxes.

Fig. 9, the potential difference between H_2 and P, due to the radial flux ψ from A'_1, is

$$e_{p_1} = \frac{\psi \log_e x_{1/a}}{2\pi k K}$$

Similarly, the potential difference between H_2 and P, due to the radial flux $-\psi$ from A'_2, is[1]

$$e_{p_2} = \frac{-\psi \log_e x_{2/b}}{2\pi k K}$$

The total potential difference between H_2 and P due to the two radial fields, and hence due to the resultant field, is the algebraic sum of the potential difference produced by each field separately:

$$e_p = e_{p_1} + e_{p_2} = \frac{\psi}{2\pi k K}\left(\log_e \frac{x_1}{a} - \log_e \frac{x_2}{b}\right)$$

But if H_2 and P are points on the same equipotential surface the potential difference between them is zero, or

$$e_p = \frac{\psi}{2\pi k K}\left(\log_e \frac{x_1}{a} - \log_e \frac{x_2}{b}\right) = 0$$

Therefore
$$\log_e \frac{x_1}{a} = \log_e \frac{x_2}{b}$$

$$\frac{x_1}{a} = \frac{x_2}{b}$$

$$\frac{x_1}{x_2} = \frac{a}{b} = \text{const.}$$

if H_2 is fixed—that is, for any one equipotential surface.

The equipotential surfaces are cylinders whose sections are circles which surround the infinitely small conductors. These circles have centers on line A_1A_2, but are not concentric with the conductors. This may be shown as follows:

Assume Cartesian axes through A'_2, the X axis containing A'_1. Let the coordinates of P be x, y,

$$x_1 = \sqrt{(A'_1A'_2 - x)^2 + y^2} = \sqrt{(a + b - x)^2 + y^2}$$
$$x_2 = \sqrt{x^2 + y^2}$$
$$\frac{x_1}{x_2} = \frac{a}{b} = \sqrt{\frac{(a + b - x)^2 + y^2}{x^2 + y^2}}$$

[1] The signs must always be properly placed. It is convenient to give ψ the sign of the point displacement under consideration. The distances from the displacement points to the points between which potential is sought should always be put in the same order in the log, as, x_1/a, x_2/b, etc. See problems, Case 11 and Case 12, Chap. X.

$$a^2x^2 + a^2y^2 = b^2a^2 + b^4 + b^2x^2 + 2ab^3 - 2ab^2x - 2b^3x + b^2y^2$$

$$(a^2 - b^2)x^2 + 2b^2(a + b)x + (a^2 - b^2)y^2 = b^2(a + b)^2$$

$$x^2 + \frac{2b^2}{a - b}x + \frac{b^4}{(a - b)^2} + y^2 = \frac{b^2(a + b)^2}{a^2 - b^2} + \frac{b^4}{(a - b)^2}$$

$$(x + \frac{b^2}{a - b})^2 + y^2 = \frac{b^2(a + b)}{a - b} + \frac{b^4}{(a - b)^2}$$

$$\left(x + \frac{b^2}{a - b}\right)^2 + y^2 = \frac{b^2(a^2 - b^2) + b^4}{(a - b)^2} = \frac{a^2b^2}{(a - b)^2}$$

$$\left(x + \frac{b^2}{a - b}\right)^2 + y^2 = \frac{a^2b^2}{(a - b)^2} \tag{7}$$

This is the equation of a circle whose center has the coordinates $\frac{b^2}{a - b}$, 0 and whose radius is $\frac{ab}{a - b}$. The circle is thus found for any given a and b.

The equipotential circle through any point P (x_P, y_P) is found as follows:

$$a = \frac{a}{a + b}S' = \frac{x_1}{x_1 + x_2}S'$$

$$b = \frac{b}{a + b}S' = \frac{x_2}{x_1 + x_2}S'$$

Substituting for a and b in (7):

$$\left\{x + \frac{\left(\frac{x_2^2}{(x_1 + x_2)^2}\right)S'^2}{S'\left(\frac{x_1 - x_2}{x_1 + x_2}\right)}\right\}^2 + y^2 = \frac{\frac{x_1^2}{(x_1 + x_2)^2}S'^2 \frac{x_2^2}{(x_1 + x_2)^2}S'^2}{S'^2\left(\frac{x_1 - x_2}{x_1 + x_2}\right)^2}$$

$$\left(x + \frac{x_2^2S'}{x_1^2 - x_2^2}\right)^2 + y^2 = \frac{x_1^2x_2^2S'^2}{(x_1^2 - x_2^2)^2}$$

$$x_1^2 = (S' - x_P)^2 + y_P^2$$

$$x_2^2 = x_P^2 + y_P^2$$

$$\left(x + \frac{(x_P^2 + y_P^2)S'}{S'^2 - 2x_PS'}\right)^2 + y^2 = \frac{\{(S' - x_P)^2 + y_P^2\}(x_P+^2 y_P^2)S'^2}{(S'^2 - 2x_P)^2}$$

$$\left(x + \frac{x_P^2 + y_P^2}{(S' - 2x_P)}\right)^2 + y^2 = \frac{\{(S' - x_P)^2 + y_P^2\}(x_P^2 + y_P^2)}{(S' - 2x_P)^2} \tag{7a}$$

The resultant lines of force are arcs of circles with centers on line n and passing through the points A'_1 and A'_2. This is shown as follows: Consider Fig. 10. The flux included in $PA'_1A'_2$ per centimeter length of cylinder due to A'_1 is $\frac{\psi \alpha_1}{2\pi}$. That in PA'_2 A'_1 due to A'_2 is $\frac{\psi \alpha_2}{2\pi}$.

The total flux between P and $A'_1 A'_2$ is the sum of these,

$$\psi_P = \frac{\psi}{2\pi} (\alpha_1 + \alpha_2)$$

The restriction that lines of force cannot cross implies that the flux between any two is constant; hence if P move along a flux line,

$$\psi_P = \frac{\psi}{2\pi} (\alpha_1 + \alpha_2) = \text{const.}$$

from which

$$\alpha_1 + \alpha_2 = \text{const.}$$

Hence:

$$\alpha = \pi - (\alpha_1 + \alpha_2) = \text{const.}$$

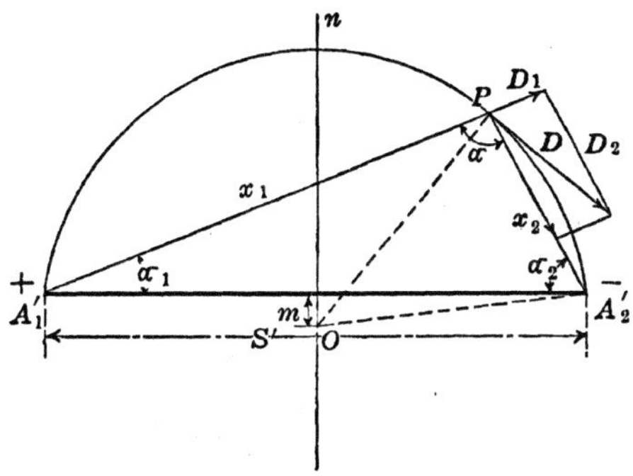

FIG. 10.

NOTE.—The equation of the line of force may also be found by writing the expression for the flux densities at points and imposing the condition that the component normal to the line of force is zero.

This condition defines a circular arc passing through A'_1 and A'_2. Choosing as before the point A'_2 as the origin of Cartesian coordinates, the equation of these circles is:

$$\left(\frac{S'}{2} - x\right)^2 + (y - m)^2 = m^2 + \left(\frac{S'}{2}\right)^2 \tag{8}$$

where m is the ordinate of the center of any particular circle. The equation of the line of force through (x, y) is found as follows:

Call the center of circle (lines of force) O

$$A_2'O = PO = \text{radius of circle}$$

$$r^2 = (A_2'O)^2 = \left(\frac{S'}{2}\right)^2 + m^2$$

$$r^2 = (PO)^2 = \left(\frac{S'}{2} - x\right)^2 + (y - m)^2$$

$$\therefore \left(\frac{S'}{2} - x\right)^2 + (y - m)^2 = m^2 + \left(\frac{S'}{2}\right)^2 \tag{8}$$

Through any point $(x_P y_P)$.

Then $$\left(\frac{S'}{2} - x_P\right)^2 + (y_P - m)^2 = \left(\frac{S'}{2}\right) + m^2$$

$$m = \frac{x_P{}^2 + y_P{}^2 - S'x_P}{2y_P}$$

Substituting this value of m in (8):

$$\left(x - \frac{S'}{2}\right)^2 + \left(y - \frac{x_P{}^2 + y_P{}^2 - S'x_P}{2y_P}\right)^2 = \left(\frac{-S'x_P\, x_P{}^2 + y_P{}^2}{2y_P}\right)^2 + \left(\frac{S'}{2}\right)^2 \tag{8a}$$

The slope of the equipotential surface at (x_P, y_P) is found from (7a).

Evaluate y in terms of x, differentiate, and put $x = x_p$.

NOTE.—Take x always +. $m-$ when below x axis.

$$\frac{dy}{dx_{es}} = -\frac{S'x_P - x_P{}^2 + y_P{}^2}{y_P(S' - 2x_P)}$$

The slope of the line of force at (x_P, y_P) is found in the same way from (8a)

$$\frac{dy}{dx_{lf}} = +\frac{y_P(S' - 2x_P)}{S'x_P - x_P{}^2 + y_P{}^2}$$

It will at once be noticed that

$$\frac{dy}{dx_{es}} = -\frac{dx}{dy_{lf}}$$

which shows that the line of force at any point is perpendicular to the equipotential surface at the same point.

The flux density, D, at any point in the resultant field is the vector sum of the flux densities due to A'_1 and A'_2 separately. At P (Fig. 10) the flux density due to A'_1 is

$$D_1 = \frac{\psi}{2\pi x_1}$$

and due to A'_2 is

$$D_2 = \frac{\psi}{2\pi x_2}$$

directed as indicated.

NOTE.—Subscript *es* refers to equipotential surface.
Subscript *lf* refers to line of force.

The triangles whose sides are x_1, x_2, S' and D_2, D_1, D may be shown to be similar, having one angle (α) equal, and the including sides proportional.

$$D_2 x_2 = D_1 x_1 = \frac{\psi}{2\pi}$$

Then
$$\frac{D}{S'} = \frac{D_1}{x_2}$$

$$D = S'\frac{D_1}{x_2} = \frac{S'}{x_2}\frac{\psi}{2\pi x_1} = \frac{S'\psi}{2\pi x_1 x_2} \qquad (9)$$

The preceding, covering infinitely small wires, is not directly applicable to the ordinary case of large parallel wires. Green's theorem, however, states that if any equipotential surface be kept at its original potential, the flux within it may be removed without any change in the external field. In Fig. 11 the circles

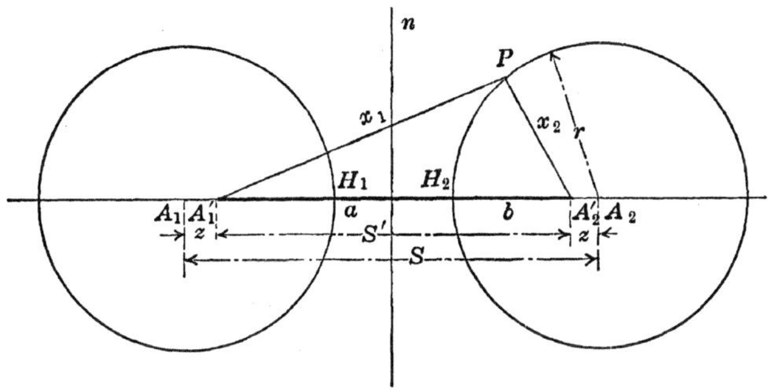

FIG. 11.

represent equipotential cylinders, surrounding flux centers A'_1 and A'_2. These cylinders may be maintained at their original potential. The interior may be filled with a conductor. This gives parallel conductors of radius r and spacing between centers S. The external field has not been changed, and the preceding discussion still applies. A_1 and A_2 must be located from A'_1 and A'_2, since r and S are the quantities given in any actual case. This is easily done:

$$a = S - r - z$$
$$b = r - z$$
$$z = \frac{b^2}{a - b} = \frac{r^2 - 2rz + z^2}{S - r - z - r + z} = \frac{r^2 - 2rz + z^2}{S - 2r}$$
$$zS = r^2 + z^2$$
$$z^2 - Sz + r^2 = 0$$
$$z = \frac{S - \sqrt{S^2 - 4r^2}}{2}$$

Since obviously z cannot be greater than $S/2$, the negative sign is taken for the radical.

$$a = S - r - z$$

$$= \frac{2S - 2r - S + \sqrt{S^2 - 4r^2}}{2}$$

$$= \frac{S - 2r + \sqrt{S^2 - 4r^2}}{2}$$

$$b = r - z$$

$$= \frac{2r - S + \sqrt{S^2 - 4r^2}}{2}$$

$$\frac{a}{b} = \frac{S - 2r + \sqrt{S^2 - 4r^2}}{2r - S + \sqrt{S^2 - 4r^2}} \tag{10}$$

$$= \frac{S - 2r + \sqrt{S^2 - 4r^2}}{2r - S + \sqrt{S^2 - 4r^2}} \times \frac{2r - S - \sqrt{S^2 - 4r^2}}{2r - S - \sqrt{S^2 - 4r^2}}$$

$$= \frac{\{- 2S^2 - 4rS + 2(S - 2r)\sqrt{S^2 - 4r^2}}{4r^2 - 4rS + 4r^2}.$$

$$= \frac{(2r - S)(S + \sqrt{S^2 - 4r^2})}{2r(2r - S)} = \frac{S}{2r} + \sqrt{\left(\frac{S}{2r}\right)^2 - 1}$$

Permittance or Capacity.—In Fig. 11 let n be a neutral plane. Represent by e_n the potential between circle H_2 and n (or H_1 and n) and by C_n the corresponding permittance to neutral per centimeter length of wires. Due to A'_1.

$$e_{n1} = \frac{\psi \log_e \frac{x_1}{S'/2}}{2\pi kK}$$

Due to A'_2.

$$e_{n2} = \frac{-\psi \log_e \frac{x_2}{S'/2}}{2\pi kK}$$

$$e_n = e_{n1} + e_{n2} = \frac{\psi}{2\pi kK}\left(\log_e \frac{x_1}{S'/2} - \log_e \frac{x_2}{S'/2}\right)$$

$$= \frac{\psi}{2\pi kK} \log_e \frac{x_1}{x_2}$$

$$C_n = \frac{\psi}{e_n} = \frac{2\pi kK}{\log_e \frac{x_1}{x_2}} = \frac{2\pi kK}{\log_e \frac{a}{b}}$$

Substituting from (10)

$$C_n = \frac{2\pi kK}{\log_e \frac{a}{b}} \tag{11}$$

$$= \frac{2\pi kK}{\log_e\left[\frac{S}{2r} + \sqrt{\left(\frac{S}{2r}\right)^2 - 1}\right]}$$

$$= \frac{5.55k10^{-13}}{\log_e\left[\frac{S}{2r} + \sqrt{\left(\frac{S}{2r}\right)^2 - 1}\right]} \text{ farads per cm.}$$

If hyperbolic tables are available, a more convenient form is

$$C_n = \frac{2\pi kK}{\cosh^{-1}\frac{S}{2r}} \tag{11a}$$

$$= \frac{5.55k10^{-13}}{\cosh^{-1}\frac{S}{2r}} \text{ farads per cm.}$$

If S/r be large, then $\sqrt{\left(\frac{S}{2r}\right)^2 - 1}$ is nearly equal to $\frac{S}{2r}$, and approximately

$$C_n = \frac{2\pi kK}{\log_e S/r} \text{ farads per cm.} \tag{11b}$$

The result corresponds exactly to the form (page 22) which would have resulted had the wires been considered very small at the start, or

$$C_n = \frac{2\pi kK}{\log_e a/b}$$

where $a = S,\ b = r$

Gradient and Flux Density.—The flux density at any point on the line joining the centers of the conductors and distant x from the inner surface of one of them is

$$D = \frac{S'\psi}{2\pi x_1 x_2} \qquad \begin{aligned} x_2 &= b + x \\ x_1 &= a - x \end{aligned} \tag{9}$$

$$= \frac{(a + b)\psi}{2\pi(ab + (a - b)x - x^2)}$$

$$= \frac{\psi\sqrt{S^2 - 4r^2}}{2\pi\left\{\frac{4rS - 8r^2}{4} + \left(\frac{2S - 4r}{2}\right)x - x^2\right\}}$$

$$= \frac{\psi\sqrt{S^2 - 4r^2}}{2\pi\{(r + x)\,(S - 2r) - x^2\}}$$

Since

$$\psi = C_n e_n$$

$$= \frac{2\pi k K e_n}{\log_e\left[\frac{S}{2r} + \sqrt{\left(\frac{S}{2r}\right)^2 - 1}\right]}$$

$$D = \frac{\sqrt{S^2 - 4r^2}}{\{2\pi(r + x)(S - 2r) - x^2\}} \quad \frac{2\pi k K e_n}{\log_e\left[\frac{S}{2r} + \sqrt{\left(\frac{S}{2r}\right) - 1}\right]}$$

$$= \frac{kKe_n\sqrt{S^2 - 4r^2}}{\{(r + x)\,(S - 2r) - x^2\}\,\log_e\left[\frac{S}{2r} + \sqrt{\left(\frac{S}{2r}\right)^2 - 1}\right]}$$

$$= \frac{0.885ke_n\sqrt{S^2 - 4r^2}10^{-13}}{\{(r+x)\,(S - 2r) - x^2\}\,\log_e\left[\frac{S}{2r} + \sqrt{\left(\frac{S}{2r}\right)^2 - 1}\right]} \text{ coulombs per cm.}^2$$

The gradient along the line of centers is

$$g = \frac{D}{kK} \tag{12}$$

$$= \frac{e_n\sqrt{S^2 - 4r^2}}{\{(r + x)\,(S - 2r) - x^2\}\,\log_e\left[\frac{S}{2r} + \sqrt{\left(\frac{S}{2r}\right)^2 - 1}\right]} \text{ kv. per cm.}$$

The gradient is greatest at the conductor surface ($x = 0$) and is

$$g_{max} = \frac{e_n\sqrt{S^2 - 4r^2}}{r(S - 2r)\log_e\left[\frac{S}{2r} + \sqrt{\left(\frac{S}{2r}\right)^2 - 1}\right]} \tag{12a}$$

$$= \frac{e_n\sqrt{\left(\frac{S}{2r}\right)^2 - 1}}{r\left(\frac{S}{2r} - 1\right)\,\log_e\left[\frac{S}{2r} + \sqrt{\left(\frac{S}{2r}\right)^2 - 1}\right]} \text{ kv. per cm.}$$

$$= \frac{e_n}{r\sqrt{\frac{S/2r - 1}{S/2r + 1}}\,\log_e\left[\frac{S}{2r} + \sqrt{\left(\frac{S}{2r}\right)^2 - 1}\right]} \text{ kv. per cm.}$$

Where hyperbolic tables are available a more convenient form is:

$$g_{max} = \frac{e_n}{r\sqrt{\frac{\frac{S}{2r} - 1}{\frac{S}{2r} + 1}}\,\cosh^{-1}\frac{s}{2r}} \text{ kv. per cm.} \tag{12b}$$

If S/r is large $\qquad g_{max} = \dfrac{e_n}{r \log_e S/r}$ (12c)

As before, this equation would have resulted had the conductors been considered small in the first place. The method of drawing the lines of force, etc., is illustrated in Chapter X.

Spheres.—In studying insulation it is sometimes convenient to use spherical electrodes. The potential between *concentric* spheres of radii R and r may be found as for concentric cylinders on page 13. The equipotential surfaces are spheres. Between two surfaces at distance x, dx centimeters apart the permittance is:

$$dC = \frac{A}{dx} kK = \frac{4\pi x^2}{dx} kK$$

$$\frac{1}{C} = \frac{1}{4\pi kK} \int \frac{dx}{x^2} = \frac{1}{4\pi kK} \left[\frac{1}{r} - \frac{1}{R} \right]$$

$$C = 4\pi kK \frac{Rr}{R - r}$$

$$e_{rR} = \frac{\psi}{C} = \frac{\psi}{4\pi kK} \frac{R - r}{Rr}$$

The potential difference due to ψ between two points distant r_1 and r_2 from the center of the sphere is

$$e_{r_1 r_2} = \frac{\psi}{4\pi kK} \frac{r_2 - r_1}{r_1 r_2}$$

The equipotential surfaces between two point electrodes or very small equal spheres may be found as follows, using Fig. 11: The difference of potential between H_2 and P due to A'_1 is

$$e_{P_1} = \frac{\psi}{4\pi kK} \left(\frac{x_1 - a}{x_1 a} \right)$$

due to A'_2 is

$$e_{P_2} = - \frac{\psi}{4\pi kK} \frac{x_2 - b}{x_2 b}$$

$$e_P = e_{P_1} + e_{P_2} = \frac{\psi}{4\pi kK} \left(\frac{x_1 - a}{x_1 a} - \frac{x_2 - b}{x_2 b} \right)$$

If H_2 and P are on the same equipotential surface

$$\frac{x_1 - a}{x_1 a} - \frac{x_2 - b}{x_2 b} = 0$$

$$\frac{x_1 x_2}{x_2 - x_1} = \frac{ab}{b - a} = \text{constant}$$

$$\frac{1}{x_1} - \frac{1}{x_2} = \text{constant}$$

is the equation for the equipotential surface

The fraction of the total flux toward P through the cone with apex at A'_1 and half angle α due to A' is[1]

$$\psi_1 = \frac{\psi}{2}(1 - \cos \alpha_1)$$

Through the cone with apex on A'_2 and half angle α_2 due to A'_2, it is

$$\psi_2 = \frac{\psi}{2}(1 - \cos \alpha_2)$$

$$\psi_p = \psi_1 + \psi_2$$

If P follows a line of force ψ_P must be constant because lines of force cannot cross.

$$\therefore \cos \alpha_1 + \cos \alpha_2 = \text{constant}$$

is the *equation of the line force.*

The equations for the gradients, etc., of two large spheres of equal radii are given below:

Spheres of Equal Size in Air (Non-grounded):[2]

$$g = \frac{e}{X} f \text{ kv. per cm.} \tag{13a}$$

where

g = gradient at surface of sphere in line joining centers.
e = volts between spheres.
X = distance between nearest surfaces in centimeters.
f = a function of X/R where R is the radius of either sphere.

Spheres of Equal Size in Air (One Sphere Grounded):

$$g = \frac{e}{X} f_1 \text{ kv. per cm.}$$

$$g = \frac{e}{X} f_0 \text{ kv. per cm.} \tag{13b}$$

where the letters have the meaning noted above, f_1 being a different function of X/R. For the case of one sphere grounded, the shanks, connecting leads, ground, etc., have a much greater effect than when both are non-grounded. For this reason the theoretical values of f_1 do not check closely the experimental results

[1] The area of a spherical surface is $4\pi R^2$. The area of a spherical sector with half angle α is $2\pi R^2(1 - \cos\alpha)$.

[2] Russel, Phil. Mag., Vol. XI, 1906.

and must not be used. Experimental values for the grounded case are given as f_0 in the accompanying table.

f and f_1 may be calculated by the following simple formulæ:[1]

$$f = \frac{X/R + 1 + \sqrt{(X/R + 1)^2 + 8}}{4}$$

$$f_1 = 1/2\left\{X/R + \sqrt{(X/R)^2 + 4}\right\}$$ (Not to be used in practice)

X/R	Calculated		Measured[2]
	f Non-grounded	f_1 Grounded (Not for use)	f_0 Grounded
0.1	1.034	1.050	1.03
0.5	1.177	1.282	1.18
1.0	1.366	1.517	1.41
2.0	1.781	2.339	1.97
3.0	2.225	3.252	2.59
4.0	2.686	4.201	3.21
6.0	3.640	6.143	....
10.0	5.600	10.091	....
15.0	8.080	15.086	....
20.0	10.580	20.081	....

The gradient at any point in the line joining the centers of the spheres and distance y from the mid-point of this line is

$$g_y = \frac{2x^2\,(x^2(f + 1) + 4y^2(f - 1))}{[x^2(f + 1) - 4y^2(f - 1)]^2} \qquad (13c)$$

If two equal spheres are never separated a greater distance than twice their radii, corona can never form, but the first evidence of overstress is spark-over. If the separation is greater than $2R$, corona forms and it is then necessary to further increase the voltage to cause spark-over. The condition for corona or spark-over will now be given. A wire in a cylinder will be taken, as the calculations are simpler and best illustrate the condition.

Condition for Spark-over and for Local Breakdown, or Corona.—For a wire in a cylinder the maximum gradient, and thus where breakdown will first occur, is at the wire surface,

$$g = \frac{e}{r \log_e R/r} \qquad (6a)$$

[1] Dean, Physical Review, Dec., 1912, April, 1913.
Dean, G. E. Review, March, 1913.

[2] Distance of grounded sphere to ground in these tests is 4 to 5 diameters.

When e is of such value that g at the wire surface just exceeds the breakdown strength of air, the air at that point becomes conducting, or corona forms, thus in effect, increasing the size of the conductor. If this increase lowers the gradient, the breakdown will be local, and we say corona is on the wire. If the ratio R/r is such that the increase in size of the conductor by the conducting air increases the gradient, the broken down area will continue to enlarge, or spark-over will occur. The condition for corona or spark-over may be found thus:

$$g = \frac{e}{r \log_e R/r}$$

For a constant value of e and R find the value of r to make g a minimum

$$1/g = x = \frac{r \log_e R/r}{e} = \frac{r}{e}(\log_e R - \log_e r)$$

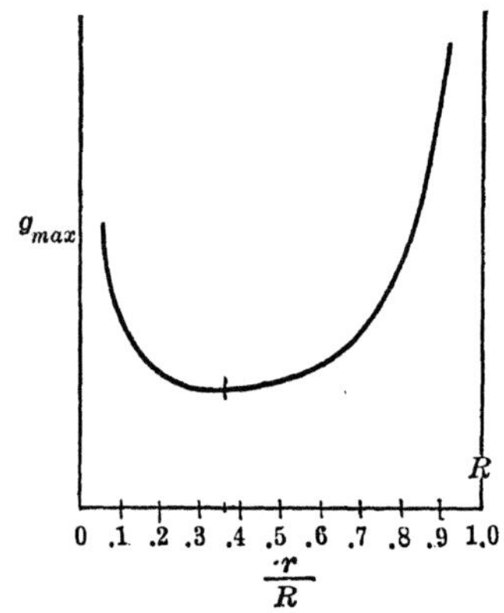

FIG. 12.—Concentric cylinders.—Variation of gradient at surface of inner cylinder as radius is changed—outer cylinder constant.

$$\frac{dx}{dr} = \frac{1}{e}(\log_e R - \log_e r - 1)$$

$$= 0 \text{ for extreme of } x$$

$$\frac{1}{e}(\log_e R - \log_e r - 1) = 0$$

$$\log_e R/r = 1$$

$$R/r = \epsilon$$

therefore

$1/g$ is maximum when $R/r = \epsilon$

or

g is minimum when $R/r = \epsilon$

In other words the stress on the air decreases with increasing r until $R/r = \epsilon$. When R/r is equal to or less than ϵ an increase in r increases g. Thus if $R/r \leqq \epsilon$ and g is brought up to the rupturing point, g progressively increases and spark-over must occur. If $R/r > \epsilon$ corona forms and the voltage must be still further increased before spark-over occurs.

This is illustrated graphically in Fig. 12. Note that this is plotted between r/R and g. Thus the minimum occurs when $r/R = 1/\epsilon$. It is interesting to note here that with a given R a cable has maximum strength when r is made such that

$$R/r = \epsilon$$

This is not the practical ratio however, as will appear later.

COLLECTED FORMULÆ FOR THE COMMON ELECTRODES

Concentric Cylinders.:

Capacity $C = \dfrac{5.55k10^{-13}}{\log_e R/r}$ farads per cm.

Gradient $g_x = \dfrac{e}{x \log_e R/r}$ kv. per cm.

Max. gradient $g = \dfrac{e}{r \log_e R/r}$ kv. per cm.

Corona does not form when

$$R/r < 2.718$$

x = distance from center of cylinder in cm.
R = cm. radius of outer cylinder.
r = cm. radius of inner cylinder.

Parallel Wires.:

Capacity or permittance to neutral

$$C_n = \frac{5.55k10^{-13}}{\log_e \left[\frac{S}{2r} + \sqrt{\left(\frac{S}{2r}\right)^2 - 1}\right]} = \frac{5.55k10^{-13}}{\cosh^{-1} \frac{S}{2r}} \text{ farads per cm.}$$

Gradient (at x cm. distance from wire surface on line through centers)

$$g_x = \frac{e_n\sqrt{S^2 - 4r^2}}{\left\{(r + x)(S - 2r) - x^2\right\} \log_e \left[\frac{S}{2r} + \sqrt{\left(\frac{S}{2r}\right)^2 - 1}\right]} \text{ kv. per cm.}$$

Max. gradient (at conductor surface)[1]

$$g = \frac{e_n\sqrt{\left(\frac{S}{2r}\right)^2 - 1}}{r\left(\frac{S}{2r} - 1\right) \log_e \left[\frac{S}{2r} + \sqrt{\left(\frac{S}{2r}\right)^2 - 1}\right]} \quad \text{or}$$

$$= \frac{e_n\sqrt{\left(\frac{S}{2r}\right)^2 - 1}}{r\left(\frac{S}{2r} - 1\right) \cosh^{-1} \frac{S}{2r}} \text{ kv. per cm.}$$

Corona does not form when

$$S/r < 5.85$$

[1] See same equations in different form, page 24. See formula 12*c*, page 25.

S = spacing between conductors centers in cm.
R = conductor radius in cm.
e_n = kv. to neutral.

Equal Spheres in Air.—Gradient (non-grounded). (At a centimeters distance from sphere surface on line joining centers.)

$$g_a = \frac{E}{X}\left\{\frac{2X^2[X^2(f+1)+4\left(\frac{X}{2}-a\right)^2(f-1)]}{[X^2(f+1)-4\left(\frac{X}{2}-a\right)^2(f-1)]^2}\right\} \text{ kv. per cm.}$$

Max. gradient (non-grounded)

$$g = \frac{e}{X} f \text{ kv. per cm.}$$

Max. gradient (one grounded)

$$g = \frac{e}{X} f_0 \text{ kv. per cm.}$$

where $$f = \frac{\frac{X}{R}+1+\sqrt{(X/R+1)^2+8}}{4}$$

and f_0 = see table, page 27.

Corona does not form when $X/R < 2.04$

R = radius of sphere in cm.
X = cm. spacing between nearest surfaces.
e = kv. bet. spheres.

Capacity or permittance

$$C = \frac{x}{36\,(f-1)}\,10^{-11} \text{ farads.}$$

Combination of Dielectrics of Different Permittivities.—When several dielectrics of different permittivities are combined, as is usually the case in practice, it becomes important to so proportion and shape the electrodes and insulations that one dielectric does not overstress another. This is of especial importance in insulators where dielectrically weak air of low permittivity is necessarily in combination with dielectrically strong insulations of high permittivity.

Dielectric Flux Refraction.—When dielectric flux lines pass from a dielectric of permittivity k_1 to another of permittivity of k_2 the lines are bent or refracted. This does not occur, of course, when the lines strike the surfaces vertically, as in a concentric cable, and between parallel plates already considered. The angle of refraction bears a definite relation to the ratio of the two permittivities and can be shown as follows:

Let AB, Fig. 13, be the common surface of the two insulations. The flux ψ_1 makes an angle ϕ_1 with the normal NN to the surface. The total flux through the equipotential surface ab is the same as the total flux through the surface cd. The voltage between a and c must be the same as the voltage between b and d, because potential at a = potential at b and potential at c = potential at d.

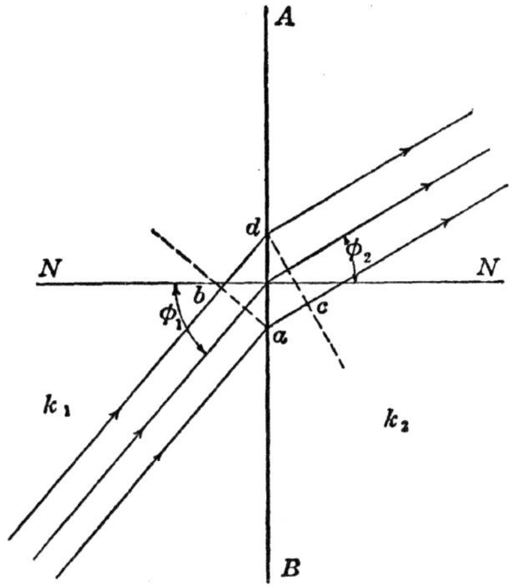

FIG. 13.—Dielectric flux refraction.

$$\psi_1 = \psi_2 = \psi$$

Therefore

$$\psi = D_1\, ab = D_2 cd \quad (14)$$

In the uniform field

$$g_1 = \frac{e}{bd} \qquad g_2 = \frac{e}{ac}$$

where e is volts, a to c = volts b to d

Therefore $$g_1 bd = g_2 ac \quad (15)$$

Combining (14) and (15)

$$\frac{D_1}{g_1}\frac{ab}{bd} = \frac{D_2}{g_2}\frac{cd}{ac}$$

But $$\frac{D_1}{g_1} = K_1 \qquad D_2/g_2 = K_2$$

Therefore $$\frac{K_1}{\tan\phi_1} = \frac{K_2}{\tan\phi_2}$$

Therefore $$\frac{\tan\phi_1}{\tan\phi_2} = \frac{K_1}{K_2}$$

Dielectric in Series.—Take the simple case of two parallel planes with two different dielectrics between them and neglect the flux concentration at the edges (Fig. 14, flux concentration not shown). As the lines are normal to the electrodes there is no refraction. The same flux passes from plate to plate.

$$\psi_1 = \psi_2 = \psi$$

$$C_1 = \frac{k_1 KA}{x_1} \qquad C_2 = \frac{k_2 KA}{x_2}$$

$$\psi_1 = C_1 e_1 = \frac{k_1 KA}{x_1} e_1$$

$$\psi_2 = C_2 e_2 = \frac{k_2 KA}{x_2} e_2$$

$$\frac{k_1 KA}{x_1} e_1 = \frac{k_2 KA}{x_2} e_2$$

$$e_1 = \frac{x_1 k_2}{x_2 k_1} e_2$$

$$e = e_1 + e_2$$

$$e = e_2 + \frac{x_1 k_2}{x_2 k_1} e_2$$

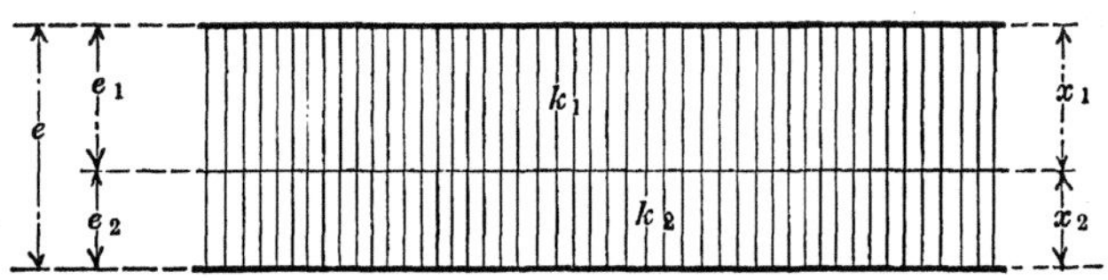

FIG. 14.—Dielectrics of different permittivities in series.

The voltages are therefore divided thus

$$e_2 = \frac{e}{\left(1 + \frac{x_1 k_2}{x_2 k_1}\right)}$$

$$e_1 = \frac{e}{\left(1 + \frac{x_2 k_1}{x_1 k_2}\right)}$$

The gradients are

$$g_1 = \frac{e_1}{x_1} = \frac{e}{x_1\left(1 + \frac{x_2 k_1}{x_1 k_2}\right)}$$

$$g_2 = \frac{e_2}{x_2} = \frac{e}{x_2\left(1 + \frac{x_1 k_2}{x_2 k_1}\right)} \qquad (16)$$

The voltages and gradients may be found in the same way for any number of insulations in series. The expression for the gradient at any point x in a combination of n insulations in series is:

$$g_x = \frac{e}{k_x\left(\frac{x_1}{k_1} + \frac{x_2}{k_2} + \dots + \frac{x_x}{k_x} + \dots + \frac{x_n}{k_n}\right)} \tag{16a}$$

Where the distance between the electrodes is greater than the radius of the edges the increase in stress at the edges becomes appreciable.

Concentric Cylinders.—For concentric cylinders the expression may be found in the same way. The flux lines in this case also are normal at every boundary surface and are hence not refracted. Consider a wire of radius r_1 surrounded by n insulations whose inside radii are respectively $r_1, r_2, \dots r_n$, and whose permittivities are $k_1, k_2, \dots k_n$.

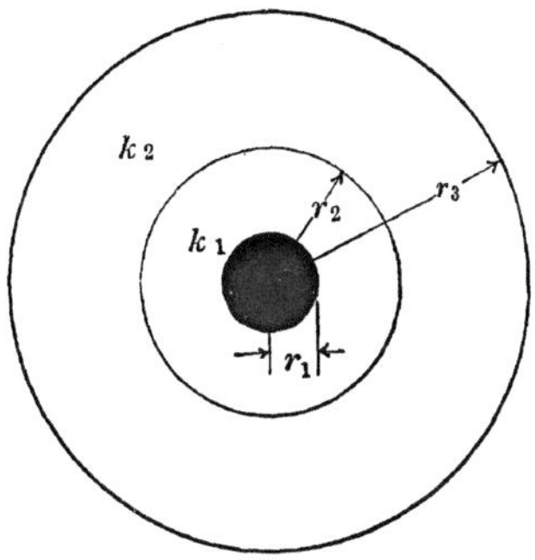

FIG. 15.

At the distance x from the center of the wire, which falls in the dielectric of inside radius r_x, outside radius r_{x+1}, and permittivity k_x, the expression for the gradient g_x is found as follows:

$$1/C = 1/C_1 + 1/C_2 + \dots + 1/C_x + \dots + 1/C_n$$

$$= \frac{1}{2\pi K}\left(\frac{\log_e r_2/r_1}{k_1} + \frac{\log_e r_3/r_2}{k_2} + \dots + \frac{\log_e (r_x + 1)/r_x}{k_x} + \dots + \frac{\log_e R/r_n}{k_n}\right)$$

$$C = 2\pi K \frac{1}{\left(\frac{\log_e r_2/r_1}{k_1} + \frac{\log_e r_3/r_2}{k_2} + \dots + \frac{\log_e r_{x+1}/r_x}{k_x} + \dots + \frac{\log_e R/r_n}{k_n}\right)}$$

$$D_x = \frac{\psi}{A}$$

$$= \frac{Ce}{A}$$

$$= \frac{2\pi K}{2\pi x} \frac{e}{\left(\frac{\log_e r_2/r_1}{k_1} + \frac{\log_e r_3/r_2}{k_2} + \dots + \frac{\log_e r_{x+1}/r_x}{k_x} + \dots + \frac{\log_e R/r_n}{k_n}\right)}$$

$$g_x = \frac{D_x}{k_x K} \tag{17}$$

$$= \frac{1}{xk_x} \frac{e}{\left(\frac{\log_e r_2/r_1}{k_1} + \frac{\log_e r_3/r_2}{k_2} + \dots + \frac{\log_e r_{x+1}/r_x}{k_x} + \dots + \frac{\log_e R/r_n}{k_n}\right)}$$

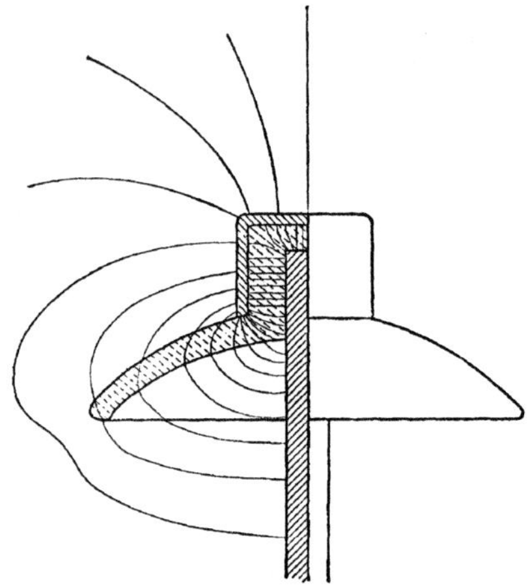

FIG. 16.—The refraction of lines of force passing through a porcelain insulator (permittivity assumed 4).

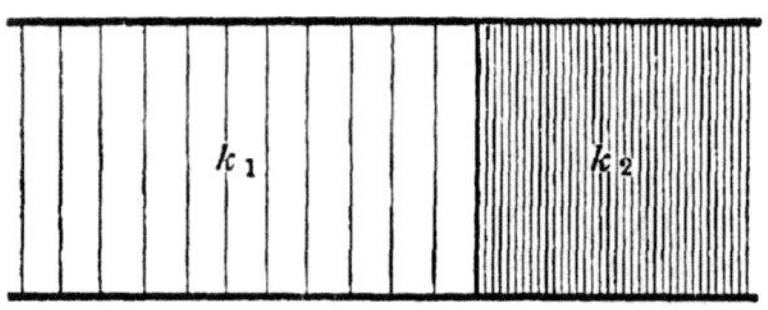

FIG. 17.—Dielectrics in multiple.

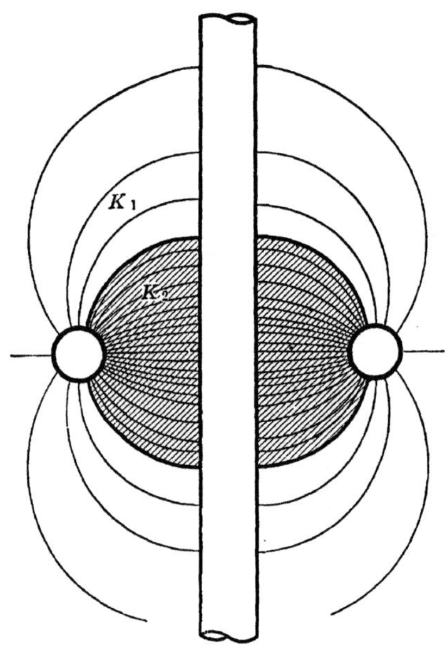

FIG. 18.—Rod and ring with two dielectrics. Boundary of dielectrics along line of force. (Not drawn to scale.)

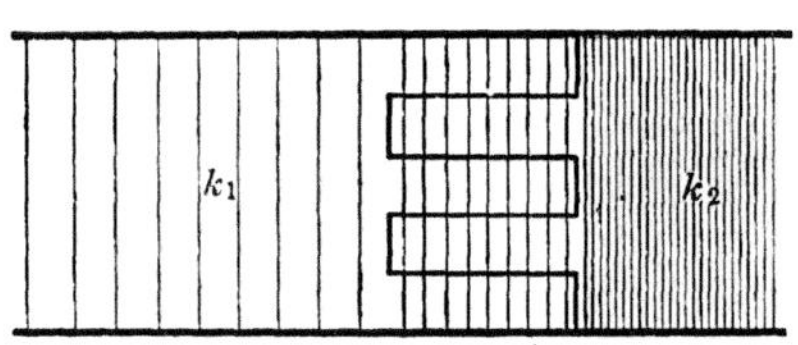

FIG. 19.—Dielectrics in multiple and in series.

Dielectrics in Multiple.—Where dielectrics are combined in multiple, the division between the dielectrics being parallel to the lines of force (Fig. 17), the stress on either is the same as it would be were the other not present. Fig. 18 shows a rod insulated from a ring by a dielectric so shaped as to make use of this fact. Where the division line is not parallel to the lines of force, some lines must pass through both dielectrics (Fig. 19). For these lines the insulators are in series, and the corresponding precautions are necessary, just as in Fig. 14.

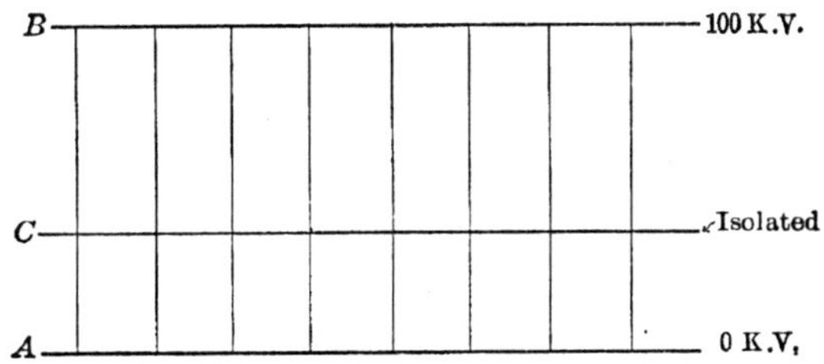

FIG. 20.—Field not changed by a thin insulated metal plate on an equipotential surface.

Flux Control.—In certain electrical apparatus it is very often possible to prevent or reduce dielectric flux concentration by superposing fields upon existing fields. When it is necessary to superpose two fields, as often happens in the course of design, it is important to see that it is properly done. For instance, as a simple case, suppose the two plates. *A* and *B* in Fig. 20, are at

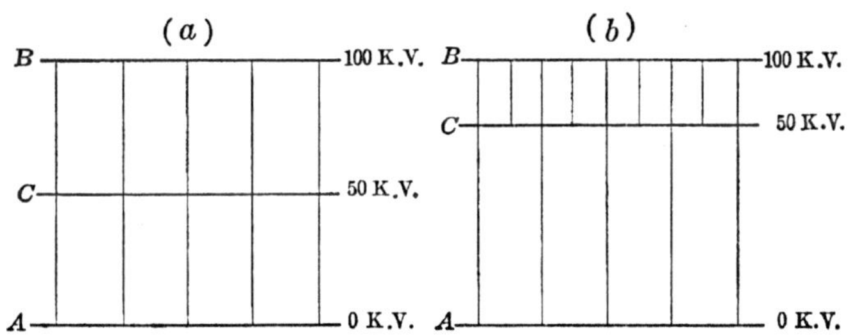

FIG. 21.—(*a*) Field not changed if the potential of the plate is the same as that of the surface upon which it rests. (*b*) Field changed by plate at potential different from the surface.

potentials of 0 and 100 kv. respectively. A thin insulated metal plate, *C*, may be placed anywhere between *A* and *B* without changing the field as long as it follows an equipotential surface—that is, parallel to *A* and *B*. Unless it follows an equipotential surface flux concentration results. If *C* is insulated and brought to a potential of 50 kv., the field will be disturbed unless *C* follows the 50 kv. equipotential surface, or, in other words, is midway between the two plates—otherwise the stress in part of the insulation will be greatly increased. See Fig. 21 (*a*) and (*b*).

Fig. 21(*a*) shows the position for no change.

Fig. 21(*b*) shows very great increase of stress on part of the insulation.

A coil of wires in which the turns are at different potentials may be placed in the field with least disturbance if the potential of each coil corresponds to the potential of the equipotential surface upon which it rests, as shown approximately in Fig. 22(*a*).

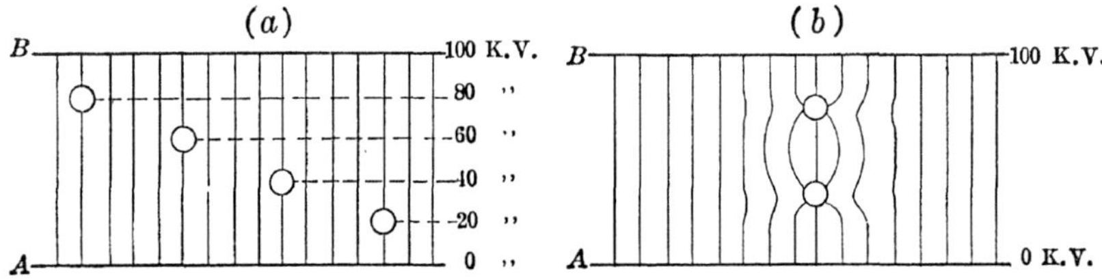

FIG. 22.—Conductors placed in a uniform field.

When insulation is used around small conductors, points, etc., the stress may be very great. This stress may be reduced by superposing a uniform field. For instance, take two small parallel wires with voltage e between them, the stress is

$$g = \frac{e}{2r \log_e \frac{S}{r}}$$

If a uniform field is superposed, as in Fig. 22(*b*) of gradient $g_1 = \frac{e}{S}$, the stress on the wires becomes

$$g_2 = 2g_1$$

These principles must be used in generator and transformer design, etc., and will be applied in a later chapter.

Imperfect Electric Elasticity.—The electric displacement has been shown to follow Hooke's law by analogy, that is.

$$\psi = Ce$$

$$D = kg\left(\frac{10^9}{4\pi v^2}\right)$$

or $$\text{strain} = \text{constant} \times \text{stress}.$$

The dielectric has so far been assumed to be perfectly elastic. For a perfectly electrically elastic material C or k must be constant and independent of the time that the stress e or g is applied. This appears to be the case for perfectly homogeneous dielectrics as air, various gases, pure oil, etc., but is not so for non-homogeneous dielectrics. As an example of imperfectly elastic dielectrics—take a cable; when potential is applied between core and case the displacement immediately reaches very nearly its full value, but gradually increases through an appreciable time slightly above its initial value. It thus *appears* that energy is slowly absorbed and this phenomenon has, therefore, been termed *absorption.* When the above condenser is disconnected from the supply, and then short circuited, the potential difference becomes zero. If the short circuit is removed a very small potential difference gradually reappears as *residual.* If such a condenser is displaced (charged), and the supply is removed, the displacement gradually disappears by conduction or *leakage.* The residual is analogous to residual stretch in an imperfectly elastic metal wire. For instance, if a steel rod is stretched and the stress is removed, it immediately assumes very nearly its initial length, but there is always very small residual stretch which very gradually disappears.

It has been shown theoretically that the phenomenon of absorption should exist for non-homogeneous dielectrics, but not for homogeneous dielectrics.

In non-homogeneous dielectrics the effect of this residual is to cause the flux to lag behind the voltage if the voltage change is rapid as in the case of high frequency. This is analogous to damping. If the change in voltage is slow, however, the effect would not result. A loop may thus be plotted (when the change of voltage is rapid) between voltage and displacement, similar to the hysteresis loop. If the frequency is very low or the dielectric is homogeneous the loop does not result.

This loop means loss, but it is not analogous to hysteresis loss in iron which is independent of time.

CHAPTER III

VISUAL CORONA

SUMMARY

Appearance.—If potential is applied between the smooth conductors of a transmission line or between concentric cylinders and gradually increased, a voltage is finally reached at which a hissing noise is heard, and if it is dark, a pale violet light can be seen to surround the conductors. This voltage is called the critical visual corona point. If a wattmeter is inserted in the line a loss is noticed. The loss increases very rapidly as the voltage is raised above this point. The glow or breakdown starts first near the conductor surface, as the dielectric flux density or gradient is greatest there. As the broken down air near the surface is conducting, the size of the conductor is, in effect, increased by conducting corona. This increases for the given voltage until the flux density or gradient is below the rupturing gradient, when it cannot spread any more. If the conductors are very close together, a spark strikes between them immediately and corona cannot form. If the conductors are far apart corona forms first, and then, if the voltage is sufficiently increased, a spark strikes across.

Whenever corona is present there is always the characteristic odor of ozone. Air consists of a mechanical mixture of oxygen (O_2) and nitrogen (N). When air is overstressed electrically the oxygen molecule is split up into O, when it becomes chemically very active. The atoms again combine by the law of probability into O = O, (O_2), the normal state, and 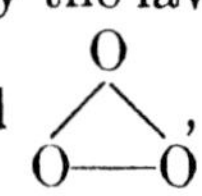, (O_3) or ozone.

Oxygen in the nascent state (O) also combines with metal, organic matter, etc., if such are present. Ozone is also not stable and is, hence, chemically active; it splits up as O_2 and O when the latter combines readily with metals, and organic matter. If the electrical stress is very high the oxygen enters into chemical combination with the nitrogen, forming oxides. The energy loss by corona is thus in a number of forms, as heat, chemical action, light, noise, convection, etc.

A.C. and D.C. Corona.—When alternating voltage higher than the critical voltage is applied between two parallel polished wires,

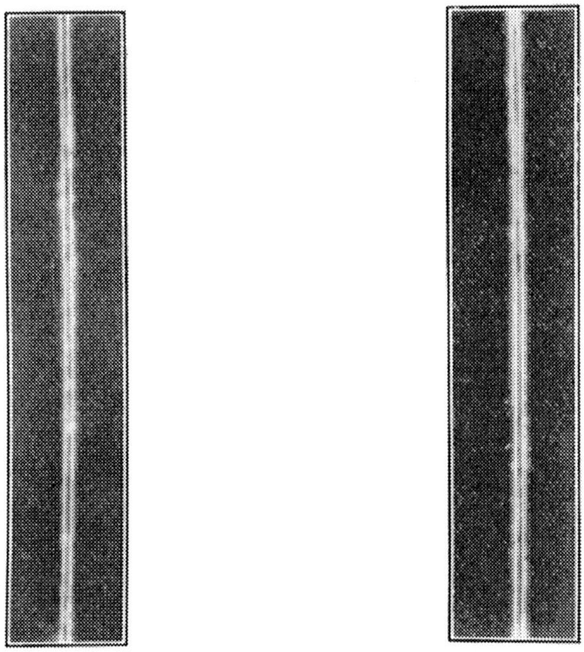

FIG. 23.—A. C. corona on polished parallel wires.

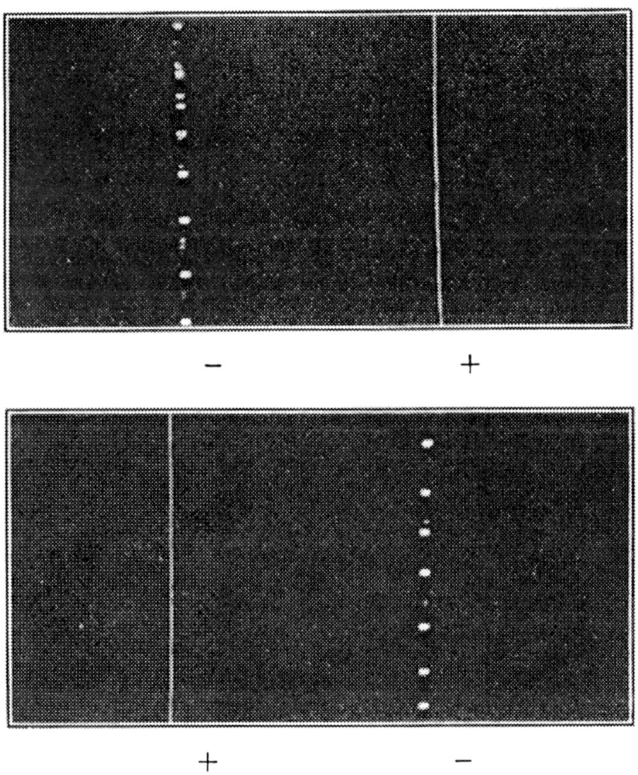

FIG. 24.—Corona on parallel wires. Iron. First polished and then operated at 120 kv. for two hours to develop spots. Diameter, 0.168 cm. Spacing, 12.7 cm. Stroboscopic photo., 80 kv.–60 v.

(*Facing page* 38.)

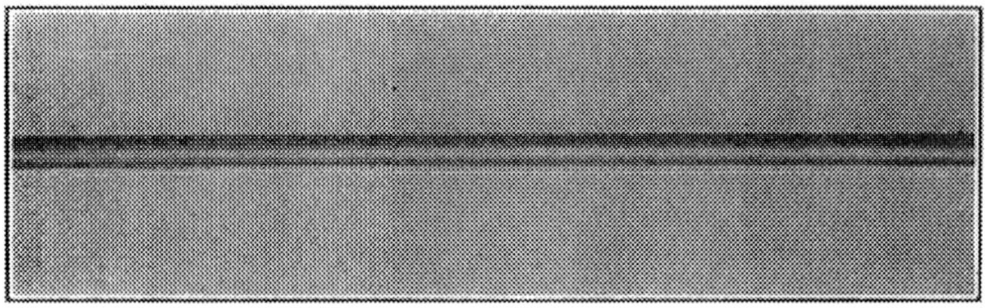

Wire "dead".

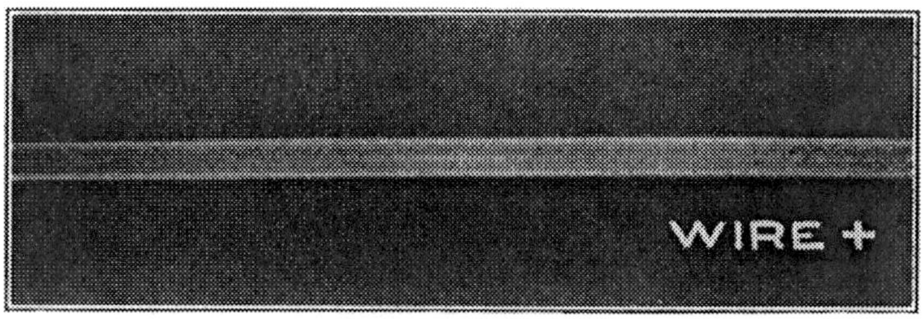

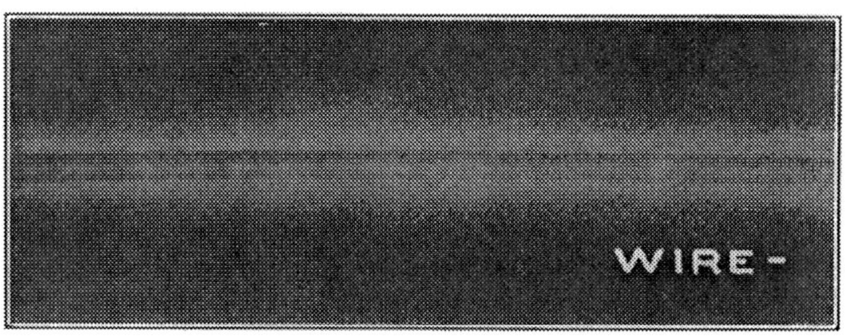

FIG. 25.—D. C. corona on smooth wires by Watson. (See Fig. 79.)

the glow is quite even, as shown in Fig. 23. After operation for a short time reddish beads or tufts form along the wire, while around the surface of the wire there is a bluish-white glow. If the conductors are examined through a stroboscope, so that one wire is always seen when at the positive half of the wave, it is noticed that the reddish tufts or beads are formed when the conductor is negative and the smoother bluish-white glow when the conductor is positive. (See Fig. 24.) A.c. corona viewed through the stroboscope has the same appearance as d.c. corona. (See Fig. 25, d.c. corona.) The d.c. corona on the + wire has exactly the same appearance as the a.c. corona on the + half of the wave; the same holds for the − wire.

Measure of Stress.—The gradient in kilovolts per centimeter is a measure of the stress on the dielectric. For parallel wires the gradient at the wire surface, and hence where the stress is a maximum and where the "dielectric elastic limit" is first exceeded, is

$$\frac{de}{dx} = g = \frac{e}{r \log_e S/r}$$

That is, if the voltage between the surface of the conductors and a point in space at infinitesimal distance dx cm. away is de, then this gradient in the limit at the surface is

$$\frac{de}{dx} = g = \frac{e}{r \log_e S/r}$$

If e is e_v, the observed voltage to neutral at which visual corona starts, g_v is a measure of the stress at breakdown. For a wire in the center of a cylinder

$$g_v = \frac{e_v}{r \log_e R/r}$$

Influence of Wire Spacing and Diameter on Apparent Strength of Air.—If the visual corona voltages are measured for a given conductor at various spacings, it is found that g_v, or the apparent strength of air, is a constant independent of the conductor spacing. It would also naturally be expected that g_v would be a constant for all conductors independent of their diameter, or in other words, air would break down under the same constant unit stress independent of the size of the conductor; just, for instance, as different sized rods of the same material would be expected to break down at the same unit stress of kilograms per square centimeter. It has long been known,[1] however, that air is *apparently* stronger at

[1] Ryan, The Conductivity of the Atmosphere at High Voltage, A.I.E.E., Feb., 1904.

the surface of small conductors than large ones. (The measured apparent strength curve is given in Fig. 28.) Of course this does not mean the voltage required to start corona is greater for small wires than for large ones (it is lower for the small conductors at a given spacing), but that the term g_v, or unit stress in the expression

$$e_v = g_v r \log_e S/r$$

is greater for air around small conductors than large ones.

This apparently greater strength of air around small conductors was long attributed to a film of condensed air at the conductor surface, because such a film would have a greater relative effect for the smaller conductors. We have given another reason for this which experiments at low air densities seem to confirm. During our first investigations we found that the relation between the apparent strength of air and the radius of the conductor could be expressed by the simple formula

$$g_v = g_o\left(1 + \frac{0.301}{\sqrt{r}}\right) \tag{18}$$

where g_o is a constant and is about 30 kv. per centimeter.[1] This means that the stress at the conductor surface at breakdown is not the same for all diameters, as already stated, but always constant at a distance $0.301\sqrt{r}$ cm. from the surface. (See Fig. 29.) This follows:

$$g_v = g_o\left(1 + \frac{0.301}{\sqrt{r}}\right) = \frac{e_v}{r \log_e S/r}$$

therefore,
$$g_o = \frac{e_v}{(r + 0.301\sqrt{r}) \log_e S/r}$$

The gradient x cm. from center of the conductor is approximately

$$g = \frac{e_v}{x \log_e S/r}$$
$$x = (r + 0.301\sqrt{r})$$

therefore, g_o is the gradient $0.301\sqrt{r}$ cm. from surface. Therefore, by substitution also

$$e_v = 30\left(1 + \frac{0.301}{\sqrt{r}}\right)r \log_e S/r \text{ max. kilovolts to neutral}$$

or for a sine wave

$$e_v = 21.2\left(1+\frac{0.301}{\sqrt{r}}\right)r \log_e S/r \text{ kilovolts effective to neutral} \tag{19}$$

[1] F. W. Peek, Law of Corona, A.I.E.E., June, 1911.

The visual corona voltage for any diameter of wire at any spacing may thus be calculated.

The explanation seems to be this: Air has a constant strength g_o for a given density, but a finite amount of energy of some form is necessary to cause rupture or to start corona. It is obvious that this definite, finite energy is necessary, as evinced by appearance of heat, that higher transient voltages are necessary, etc. This will be more fully discussed later. Hence the stress at the conductor surface must exceed the elastic limit g_o, or be increased to g_v in order to supply the necessary rupturing energy between the conductor surface and finite radial distance in space away ($0.301\sqrt{r}$ cm.) where the stress is g_o, and breakdown occurs.

Application of the Electron Theory.—The electron theory may also be applied in agreement with the above. Briefly:

When low potential is applied between two conductors any free ions in the field are set in motion. As the potential and, therefore, the field intensity or gradient is increased the velocities of the ions increase. At a gradient of $g_o = 30$ kv./cm. ($\delta = 1$) the velocity of the ions becomes sufficiently great over the mean free path to form other ions by collision with molecules. This is the property of the negative ion or electron. This gradient is constant and is called the dielectric strength of air. When ionic saturation is reached at any point the air becomes conducting, and glows, or there is corona or spark.

Applying this to a wire in the center of a cylinder: When a gradient g_v is reached at the wire surface any free ions are accelerated and produce other ions by collision with atoms or molecules, which are in turn accelerated. The ionic density is thus gradually increased by successive collisions until at $0.301\sqrt{r}$ cm. from the wire surface, where $g_o = 30$, ionic saturation is reached, or corona starts. The distance $0.301\sqrt{r}$ cm. is of course many times greater than the mean free path of the ion and many collisions must take place in this distance. Thus, for the wire, corona cannot form when a gradient of g_o is reached at the surface, as at any distance from the surface the gradient is less than g_o. The gradient at the surface must, therefore, be increased to g_v so that the gradient a finite distance away from the surface, $0.301\sqrt{r}$ cm., is g_o. This is the same as saying that energy is necessary to start corona, and this energy is the $\Sigma\frac{mv^2}{2}$ of the energy of the moving ions necessary to produce ionic saturation.

Very Small Spacings or Films.—If conductors are placed closer together than this necessary free accelerating or energy storage distance, $0.301\sqrt{r}$ cm., the rupturing force or gradient must be increased. This will be better illustrated by later experiments, in which at small spacings air has been made to withstand gradients as high as 200 kv. per centimeter.

Air Density.—Thus far the discussion has been limited to air at a constant density, or, in other words, constant pressure and temperature. The above energy explanation can be still further checked by considering air at different densities. The value of g_o given is for air density at sea level (25 deg. C., 76 cm. barometer). This has been taken as the standard, or for the density factor $\delta = 1$. Air at other densities, due to change in temperature and barometric pressure, is expressed as a fraction of this. The relative density for any temperature or pressure is

$$\delta = \frac{3.92b}{273 + t}$$

For instance, if the temperature is kept at 25 deg. C. and pressure is reduced to 38 cm.

$$\delta = \frac{3.92 \times 38}{273 + 25} = 0.50$$

or 1/2 atmosphere. As the air density, or δ is decreased, the air is less able to resist the electric stress at the increased molecular spacing. Theoretically the strength of the air g_o in bulk between parallel planes should decrease directly with δ.

$$g_o = 30\delta$$

g_v, however, the apparent strength of air in a non-uniform field, if the energy theory is true, cannot decrease directly with δ: the energy storage distance should be $0.301\sqrt{r}\phi\ (\delta)$, or the complete expression should take the form

$$g_v = g_o\delta\left(1 + \frac{0.301}{\phi(\delta)\ \sqrt{r}}\right)$$

We have found experimentally that the energy storage distance[1]

$$a = 0.301\ \frac{\sqrt{r}}{\sqrt{\delta}}\ \text{cm.; that is,}\ g_v = 30\delta\left(1 + \frac{0.301}{\sqrt{\delta r}}\right)$$

The electron theory may also be applied here:

[1] This formula may be used to determine the strength of compressed air.

g_o, the strength of air, varies directly with δ. g_v, however, cannot vary directly with δ because with the greater molecular spacing or mean free path of the ions at lower air densities, a greater "accelerating" distance is necessary in the equation $a = 0.301\sqrt{r/\delta}$ that is, "a" increases with decreasing δ. This is shown experimentally by sphere gap tests.

For parallel planes $r = \infty$

$\therefore g_v = g_o\delta$, as expected.

The equation for the visual critical corona voltage may now be written:

$$e_v = g_v\delta r \log_e S/r = g_o\delta\left(1 + \frac{0.301}{\sqrt{\delta r}}\right)r \log_e S/r \text{ kv.} \qquad (20)$$

where e_v is Kv to neutral. $g_o = 30$ for max., $g_o = 21.1$ for effective sine wave.

It may be of interest to note that with a pair of smooth wires, of a known diameter, high voltages may be accurately measured anywhere by noting the spacing at which corona starts, the temperature and barometric pressure, and substituting in the above formula.

The various other factors affecting visual corona will be mentioned.

Conductor Surface—Cables—Material.—For rough or weathered conductors corona starts at lower voltages. This is taken care of by the irregularity factor, m_v. For cables and weathered wires the complete formula becomes

$$e_v = m_v\delta g_v \, r \log_e S/r$$

For the same surface condition the starting point is independent of the conductor material.

Oil and Water, Current in the Conductor, Wave Shape, Etc.—Water, sleet and snow lower the visual corona voltage.

Oil on the conductor surface has very little effect.

Humidity has no effect upon the starting point of visual corona.

Initial ionization over a considerable range has no appreciable effect at commercial frequencies.

Current in the wire has no effect except that due to heating of the conductor and surrounding air.

Wave Shape.—The corona point at commercial frequencies depends upon the maximum value of the wave. When results are given in effective volts a sine wave is assumed. With peaked wave, corona starts at a lower effective voltage than with a flat-top wave. D.c. corona starts at a value corresponding to the maximum of the a.c. wave or 41 per cent. higher than the effective a.c. critical voltage.

The above summary will now be taken up more in detail and experimental proof given.

Experimental Study of Visual Corona

Effect of Spacing and Size of Conductor.—The visual critical voltage, or voltage at which visual corona starts on polished conductors of a given diameter and spacing, at constant air density, can be repeatedly checked within a small per cent. Visual tests were made on two parallel polished conductors supported indoors on wooden wheels in a wooden framework. The wires were not allowed to come directly in contact with the wood but rested on aluminum shields. (See Fig. 26.) The object of the shields was to prevent the glow at low voltages which would take place if the wires came in contact with the wood. The tests were made in a dark room, and method of procedure was as follows: Conductors of a given size were placed upon a framework; voltage was applied and gradually increased until the visual critical corona point was reached. Critical points were taken at various spacings and recorded as in Table I.

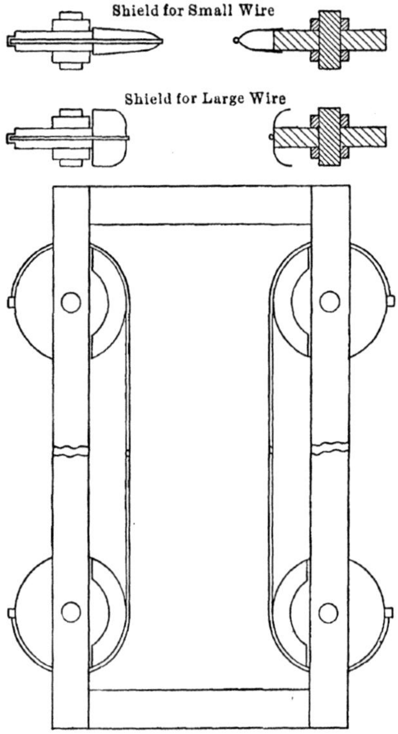

FIG. 26.—Wire support for visual corona tests.

As the visual critical voltage, e_v, is the voltage at breakdown of the air, the surface gradient corresponding to this voltage must be the stress at which air ruptures. This is called the visual critical gradient g_v. Where the wires are far apart or S/r is large

$$\frac{de}{dr} = g_v = \frac{e_v}{r \log_e S/r}$$

where

e_v = the (maximum) voltage to neutral.
r = radius of the conductor in cm.
S = distance between centers of conductors in cm.

Values of gradient calculated for a given conductor at various spacings are given in Table I. (See Fig. 27.) It is seen that the breakdown gradient is constant, or independent of the spacing. This test was repeated for various diameters. The values of the gradients are tabulated in Table II and plotted in Fig. 28.

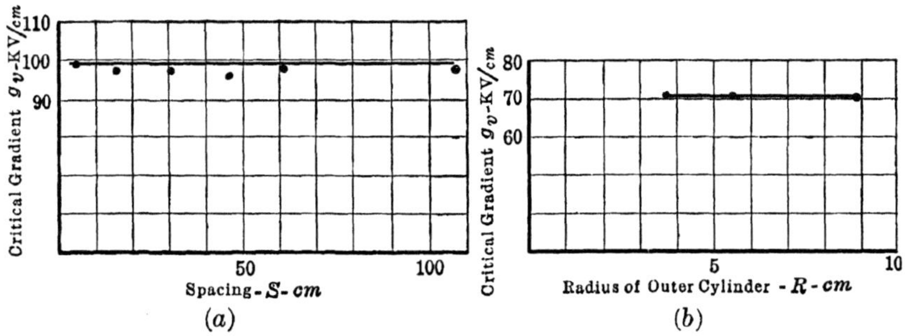

FIG. 27(a).—Variation of visual critical gradient with spacing. (Parallel wires. Diameter constant = 0.034 cm.)

FIG. 27(b).—Variation of visual critical gradient with radius of outer cylinder. (Concentric cylinders. Diameter of inner wire constant = 0.118 cm.)

TABLE I.—VISUAL CRITICAL VOLTAGES AND g_v WITH VARYING SPACING AND CONSTANT DIAMETER

(Polished Parallel Copper Conductors—Diameter 0.0343 cm.)

s cm.	e_v^1 kilovolts between conductors (effective)	e_v^1 kilovolts between conductors (maximum)	g_v kv./cm. (maximum)
2.54	12.1	17.0	99.5
2.93	12.4	17.4	99.0
3.18	12.5	17.7	98.5
3.81	13.0	18.4	99.0
4.45	13.5	19.0	99.5
5.08	13.8	19.4	100.0
5.73	14.0	19.8	99.0
7.62	14.5	20.5	99.0
15. 2	16.0	22.6	97.2
30. 5	17.7	25.0	97.2
45. 6	18.7	26.3	96.1
61. 0	19.4	27.4	98.0
106. 8	20.6	29.0	97.2
		Average,	99.0

TABLE II.—VARIATION OF g_v WITH DIAMETER OF CONDUCTORS
(Average Values for Polished Parallel Wire. Corrected to 25 deg. C. 76 cm. Barometric Pressure)

Diameter, cm.	$\frac{de}{dr} = g_v$ kv./cm. (maximum)	Material	Diameter, cm.	$\frac{de}{dr} = g_v$ kv./cm. (maximum)	Material
0.0196	116	Tungsten	0.2043	59.0	Copper
0.0343	99	Copper	0.2560	57.0	Aluminum
0.0351	94	Copper	0.3200	54.0	Copper
0.0508	84	Aluminum	0.3230	50.5	Copper
0.0577	82	Aluminum	0.5130	49.0	Copper
0.0635	81	Tungsten	0.5180	46.0	Copper
0.0780	76	Copper	0.6550	44.0	Copper
0.0813	74	Copper	0.8260	42.5	Copper
0.1637	64	Copper	0.9280	41.0	Copper
0.1660	64	Iron			

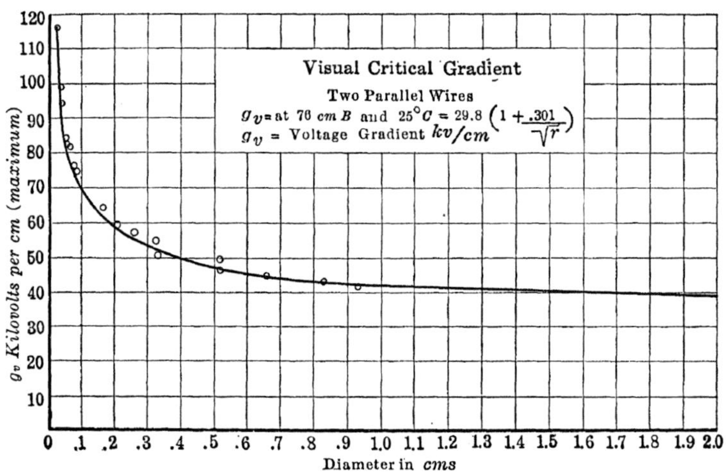

FIG. 28.—Variation of visual critical gradient with size of wire.

The gradient at breakdown at the conductor surface is not constant with varying diameters, but increases with decreasing diameters of conductors—in other words, air is apparently stronger at the surface of small wires than large ones.

The apparent increase in the dielectric strength of air surrounding small conductors was explained some years ago as due to a condensed air film at the surface of the conductor. If this were so, a higher critical gradient would be expected for tungsten than for aluminum. That is, the air film should be denser around the denser metals. These experiments show that the gradient is not affected by the material or density of the conductor. Ryan[1] has also explained the apparent increase by the electron theory.

The explanation first offered in Law of Corona I,[2] and already outlined in this chapter, will now be given more in detail: Assume that air at a given density has a constant strength g_o, but that a finite amount of energy, be this energy the $\Sigma \frac{mv^2}{2}$ of the moving ions or whatever form it may, is necessary to cause rupture or start corona. Then the stress at the conductor surface must exceed the elastic limit g_o, or be increased to g_v in order to supply the rupturing energy between the conductor surface and a finite radial distance in space where the stress is g_o and breakdown occurs, or, in other words, a finite thickness of insulation must be under a stress equal to, or more than g_o.

Just before rupture occurs the gradient at the conductor surface is

$$g_v = \frac{e_v}{r \log_e S/r} \tag{21}$$

the gradient a cm. away from the surface is

$$g_o = \frac{e_v}{(r + a) \log_e S/r} \tag{22}$$

Theoretically, one is also led to believe

$$a = \phi(r)$$

or

$$g_o = \frac{e_v}{(r + \phi(r)) \log_e S/r} \tag{23}$$

It now remains to test this out experimentally and find $\phi(r)$.

The equation of the experimental curve is found to be:

$$g_v = g_o\,(1 + \alpha/\sqrt{r}) = 29.8\left(1 + \frac{0.301}{\sqrt{r}}\right) \tag{24}$$

Substituting (24) in (21)

$$g_o\left(1 + \frac{0.301}{\sqrt{r}}\right) = \frac{e_v}{r \log_e S/r}$$

$$g_o = \frac{e_v}{(r + 0.301\sqrt{r}) \log_e S/r}$$

Thus the experimental values bear out the above theory

$$a = \phi(r) = 0.301\sqrt{r}$$
$$g = 29.8 = \text{constant}$$

[1] A.I.E.E., January, 1911. See also papers by Whitehead, A.I.E.E., June, 1910, 1911, etc.

[2] A.I.E.E., June, 1911.

That is, at rupture the gradient a finite distance away from the conductor surface, which is a definite function of r, is always constant. (See Fig. 29.)

$$g_v = 29.8\left(1 + \frac{0.301}{\sqrt{r}}\right) \text{ kv. per cm.}$$

A similar investigation made on wires in the centers of metal cylinders (see Figs. 27 and 30) shows as above that the visual

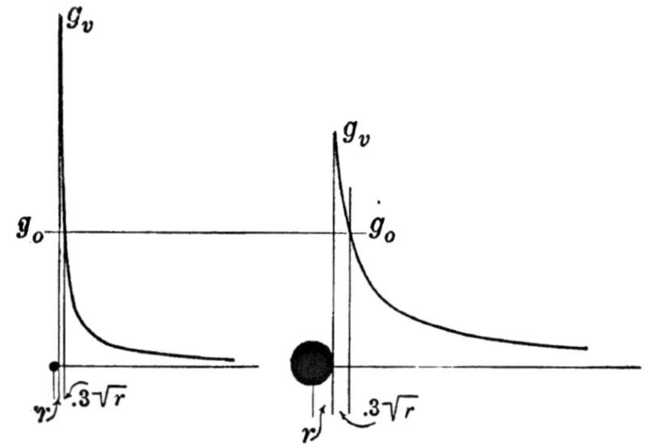

FIG. 29.—g_v and g_o for small and large wires.

critical gradient, g_v, increases with decreasing radius r, of the wire, but is independent of the radius R, of the outer cylinder. The relation between r and g_v is given in Table V. For a wire in the center of a cylinder the gradient g_o is slightly higher than for similar parallel wires, apparently due to the fact that the field is everywhere balanced for a wire in the center of a

FIG. 30.—Apparatus for determining the visual corona voltage in concentric cylinders.

cylinder, which is not the case for parallel wires. The expression for the apparent strength of air for a wire in the center of a cylinder is

$$g_v = 31\left(1 + \frac{0.308}{\sqrt{r}}\right) \text{ kv. per cm.}$$

It thus seems that $g_c = 31$ is the true dielectric strength of air.

The method of reducing the results to equations was as follows: Various functions of r and g_v were plotted for parallel wires from Table II, until it was found that a straight-line law obtained between g_v and $1/\sqrt{r}$. All values of g_v and $1/\sqrt{r}$ were then tabulated as in Table III and plotted as in curve (Fig. 31). Points

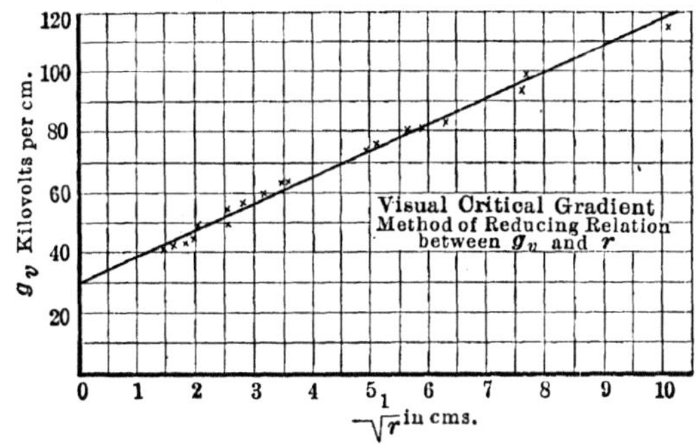

FIG. 31.—Relation between g_v and $\frac{1}{\sqrt{r}}$.

TABLE III.—RELATION OF VISUAL CRITICAL VOLTAGE GRADIENT TO RADIUS

(Experimental Values Corrected to 76 cm. Barometer and 25 deg. C. Parallel Wires)

Diameter, cm.	g_v kv./cm. (max.)	Radius, r = cm.	$\frac{1}{\sqrt{r}}$ cm.	Diameter, cm.	g_v kv./cm. (max.)	Radius, r = cm.	$\frac{1}{\sqrt{r}}$ cm.
0.0196	116.0	0.0098	10.10	0.202	59.1	0.101	3.13
0.0343	99.0	0.0172	7.65	0.257	56.7	0.128	2.76
0.0350	94.0	0.0175	7.58	0.320	54.3	0.160	2.51
0.0508	84.0	0.0254	6.27	0.322	49.6	0.161	2.51
0.0577	81.5	0.0288	5.90	0.513	48.8	0.256	2.01
0.0635	81.0	0.0317	5.64	0.518	44.5	0.259	1.94
0.078	76.0	0.0390	5.08	0.655	43.7	0.327	1.82
0.0813	73.5	0.0406	4.96	0.826	42.2	0.413	1.57
0.1635	63.8	0.0818	3.51	0.928	40.6	0.464	1.44
0.1660	63.4	0.0830	3.45				

varying widely from the average straight line were then discarded as probably in experimental error. The remaining points were then divided into two equal groups and tabulated as in Table IV.

TABLE IV.—RELATION OF g_v AND $\frac{1}{\sqrt{r}}$

(Showing ΣΔ Reduction)

g_v	$\frac{1}{\sqrt{r}}$	g_v	$\frac{1}{\sqrt{r}}$
99	7.65	59.0	3.13
82	5.90	54.0	2.51
81	5.64	50.5	2.51
76	5.08	49.0	2.01
74	4.96	41.0	1.44
Σ412	Σ29.23	Σ253.5	Σ11.60

$$\Delta\Sigma g_v = 158.5 \qquad \Delta\Sigma \frac{1}{\sqrt{r}} = 17.63$$

$$\Sigma\Sigma g_v = 665.5 \qquad \Sigma\Sigma \frac{1}{\sqrt{r}} = 40.83$$

$$c_1 = \frac{158.5}{17.63} = 9.00$$

$$g = \frac{665.5 - (9 \times 40.8)}{10} = 29.8$$

Therefore:

$$g_v = 29.8 + \frac{9}{\sqrt{r}}$$

$$= 29.8\left(1 + \frac{0.301}{\sqrt{r}}\right)$$

In order to give proper weight to the points the ΣΔ method was used in the evaluations of the constants for the equation above. This method of reduction, which is self explanatory, is very convenient and especially suitable where a large number of experimental points have been obtained, in which case the results are as reliable as, or more so, than when few points are taken and the unwieldy method of least squares used.

Temperature and Barometric Pressure.—The density of the air varies directly with the pressure, and inversely as the absolute temperature. In these investigations the air density at a temperature of 25 deg. C. and a barometric pressure of 76 cm. has been taken as standard. If the air density at this temperature and pressure is taken as unity, the relative density at other temperatures and pressures may be expressed in terms of it, thus:

$$w = \frac{0.004656}{273 + t}$$

where w = the weight of air in grams per cubic centimeter
b = barometric pressure in centimeters
t = temperature in degrees centigrade.

At 25 deg. C. and 76 cm. pressure

$$w_{25^\circ - 76\text{cm.}} = \frac{0.00465 \times 76}{273 + 25} = 0.001185 \text{ grams}$$

w at any other temperature and pressure is

$$w_{tb} = \frac{0.00465b}{273 + t}$$

$$\frac{w_{tb}}{w_{25^\circ\ 76\text{cm.}}} = \frac{0.00465b}{(273 + t)0.001185} = \frac{3.92b}{273 + t} = \delta$$

$$\delta = \frac{3.92b}{(273 + t)}$$

On the theory that definite energy is necessary to start disruption or glow, g_o should vary directly with the air density factor δ; g_v, however, should not vary directly with δ, as the thickness of the energy or ionizing film should also be a function of δ. The equation for g_v should take the form

$$g_v = g_o\left(1 + \frac{\alpha}{\phi(\delta)\sqrt{r}}\right) \tag{25}$$

Whether δ is varied by change of temperature or air pressure the effect should be the same as long as the temperature is not so high that the air is changed or affected by the heat, as ionization, etc.

Temperature.—A series of experiments on visual corona was carried on over a temperature range of −20 deg. C. to 140 deg. C. The apparatus is shown in Fig. 30. It consists of a polished wire in the center of a brass cylinder. The cylinder was placed horizontally in a large asbestos lined "hot box," heated by grids at the bottom. In order to get uniform temperature the cylinder was shielded, and time was allowed to elapse after each reading. Temperature was observed by a number of thermometers distributed in the hot box.

TABLE V.—RELATION OF VISUAL CRITICAL VOLTAGE GRADIENT TO RADIUS (76 cm. Barometer—25 deg. C.—Concentric Cylinders)

Radius r cm.	a.c. g_v kv./cm. (max.)	R cm.	d.c.[1] g_v kv./cm.	r cm.	a.c. g_v kv./cm. (max.)	R cm.	d.c. g_v kv./cm.
0.059	70.4	8.89—5.55—3.65	69.0	0.327	48.1	8.89-5.55-3.65	42.0
0.103	60.7		59.5	0.476	44.9		39.0
0.127	58.4		55.5	0.794	41.9		......
0.129	56.6		54.5	0.953	41.2		......
0.190	52.7		49.5	1.113	39.7	3.81	......
0.199	52.7		48.5	1.270	39.2	3.81	......
0.206	51.6		47.5	1.588	38.4	3.81	......
0.254	49.9		44.5	1.905	37.8	3.81	......
0.318	47.1		43.0	2.540	35.0	......	......

After the heating became uniform, voltage was applied and gradually increased until the glow appeared. The central conductor was observed through a window placed in the front part of the box so that the whole length of the conductor could be seen. It was found that it made no appreciable difference in the starting voltage whether or not the box and tube were "aired out" after each test.

Three sizes of brass cylinders were used having inside radii of 8.89, 5.55, and 3.65 cm. respectively. The central conductors ranged in size from 0.059 to 0.953 cm. radii. Tables VI and VII are typical data tables.

TABLE VI.—VARIATION OF STRENGTH OF AIR WITH TEMPERATURE
For Polished Copper Tube Inside of Brass Cylinder
$r = 0.953$ $R = 5.55$ cm.

Observed values					Calculated from equation
Kv. effective	Temp. C°	b cm.	δ	g_v(max.)	g'_v(max.)
48.5	18	75.4	1.016	40.7	41.4
46.5	37	75.4	0.954	39.1	39.1
45.2	50	75.4	0.915	38.0	37.7
43.4	66	75.4	0.873	36.5	36.2
41.0	85	75.4	0.826	34.5	34.5
39.6	100	75.4	0.793	33.3	33.3
37.6	119	75.4	0.754	31.6	31.9

[1] D.c. values from Watson, Jour. I. E. E., June, 1910, Fig. 21.

TABLE VII.—VARIATION OF STRENGTH OF AIR WITH TEMPERATURE

For Polished Copper Tube Inside of Brass Cylinder

$r = 0.476$ cm. $R = 5.55$ cm.

Observed values					Calculated from equation
Kv. effective	t	b	δ	g_v(max.)	g'_v(max.)
41.0	−13	75.5	1.139	49.6	50.0
40.0	0	75.5	1.084	48.3	48.0
37.0	20	74.9	1.001	44.8	44.9
35.7	41	75.5	0.942	43.2	42.7
33.2	70	75.5	0.863	40.1	39.7
31.5	87	75.5	0.823	38.1	38.1
29.5	121	75.5	0.753	35.7	35.4
28.7	130	75.5	0.734	34.7	34.7

Columns 1, 2 and 3 give observed values. The gradient at the surface of the inner cylinder is

$$g = \frac{e}{r \log_e R/r}$$

Column 5 gives the surface gradient for the voltage, e_v, calculated directly from observed values. It can be seen from the data, that g_v for a given r, varies with δ, but is independent of R or S.

By $\Sigma\Delta$ reduction of all of the data the following equations connecting g_v with r and δ were obtained:

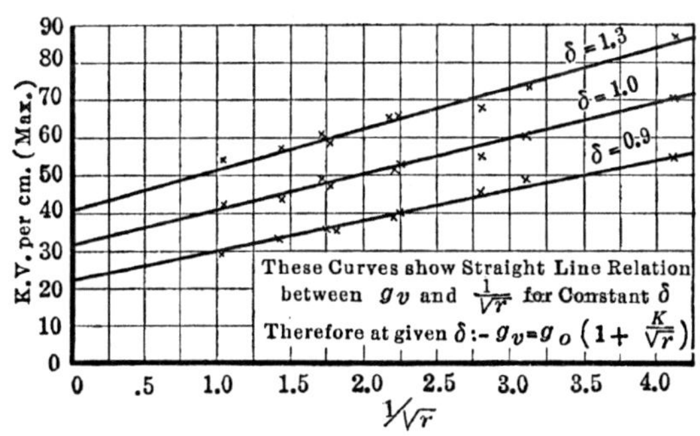

FIG. 32.—Effect of temperature on the strength of air. × measured values.

For Concentric Cylinders.

$$g_v = 31\delta\left(1 + \frac{0.308}{\sqrt{\delta r}}\right) \text{ kv. per cm. max.} \qquad (25a)$$

For Parallel Wires.

$$g_v = 29.8\delta\left(1 + \frac{0.301}{\sqrt{\delta r}}\right) \text{ kv. per cm. max.} \qquad (25b)$$

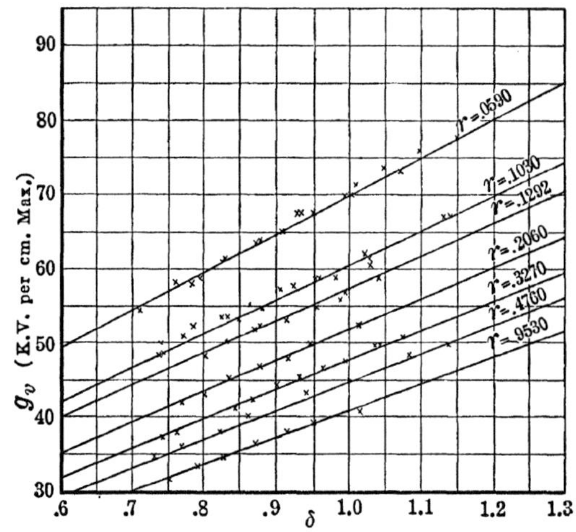

FIG. 33.—Effect of temperature on the strength of air. (r = radius of wire in cm. × measured values. Drawn curves calculated.)

Referring to the tables, column 6 gives values of g_v calculated from equation (25a). By comparing with the experimental values in column 5 it is seen that the difference is generally less than 1 per cent.

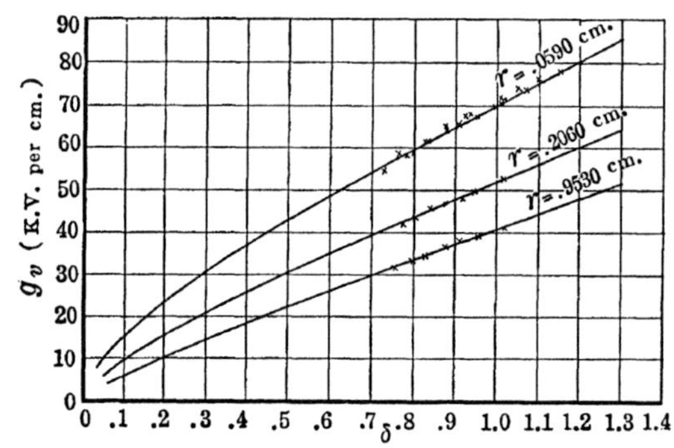

FIG. 34.—Effect of temperature on the strength of air. (× measured values. Drawn curves calculated.)

g_o has a slightly higher value for wires in a concentric cylinder than for parallel wires. This does not mean that the strength of air differs in the two cases. For a wire in a cylinder the field is balanced all around and should give more nearly the true value.

In Figs. 32, 33 and 34 the drawn lines are the calculated values, while the crosses are the observed values.

Barometric Pressure.—It will be noted in Fig. 34 that while the calculated curve is almost a straight line down to $\delta = 0.5$, below this point there is a decided bend to zero. The lower part of this curve was drawn from calculations. In order to check experimentally the above law over a wide range of δ, and also to show that the effect was the same whether the change was made by varying temperature or pressure, tests were made over a large pressure range.

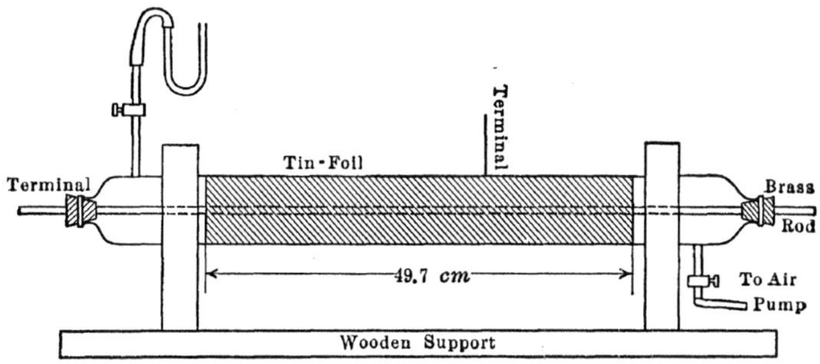

FIG. 35.—Apparatus for determining the effect of pressure on strength of air.

A glass cylinder lined with tin foil 7.36 cm. in diameter with a small slit window in the center was used for this purpose. (See Fig. 35.) Tests were made on wires 0.508 cm. to 0.157 cm. in diameter, and

TABLE VIII.—VARIATION OF STRENGTH OF AIR WITH PRESSURE
Diameter of Brass Rod = 0.381 cm. in 7.36-cm. Diameter Glass Tube Covered with Tin Foil

Pres. abs. cm. Hg	Volts read (eff.)	Temp. deg. C.	$\delta = \frac{3.92b}{273 + t}$	$g_v = \frac{e_v}{r(\log_e R/r)}$ kv./cm. (eff.)	g_v max., measured kv./cm.	g_v max., calculated $31\delta\left(1 + \frac{0.301}{\sqrt{\delta r}}\right)$ kv./cm.
5.3	2,880	27.0	0.069	5.08	7.19	7.80
10.7	4,580	24.0	0.141	8.09	11.45	12.40
11.2	4,920	27.0	0.146	8.68	12.28	12.15
19.3	7,400	27.0	0.252	13.06	18.47	18.58
27.7	9,550	27.0	0.362	16.87	23.85	24.06
36.6	12,000	27.0	0.478	21.20	30.00	29.60
46.4	14,450	25.0	0.612	25.50	36.07	35.70
47.0	14,600	27.0	0.614	25.80	36.50	35.80
55.7	16,300	27.0	0.728	28.80	40.75	40.75
60.0	17,760	25.5	0.792	31.35	44.30	43.50
66.0	18,400	27.0	0.867	32.50	46.00	46.75
75.0	21,100	25.0	0.997	37.25	52.70	52.25

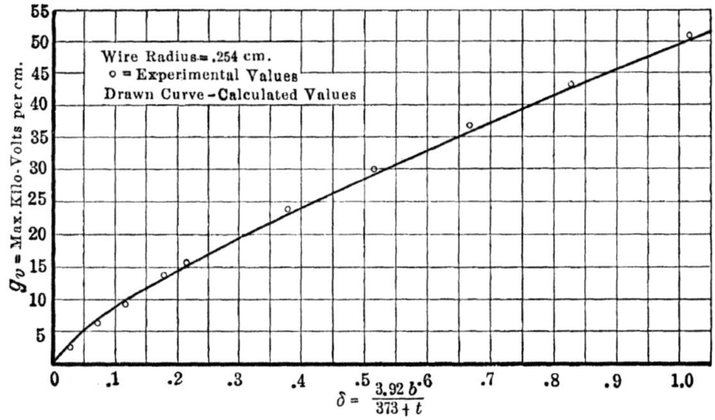

FIG. 36.—Effect of pressure on the strength of air.

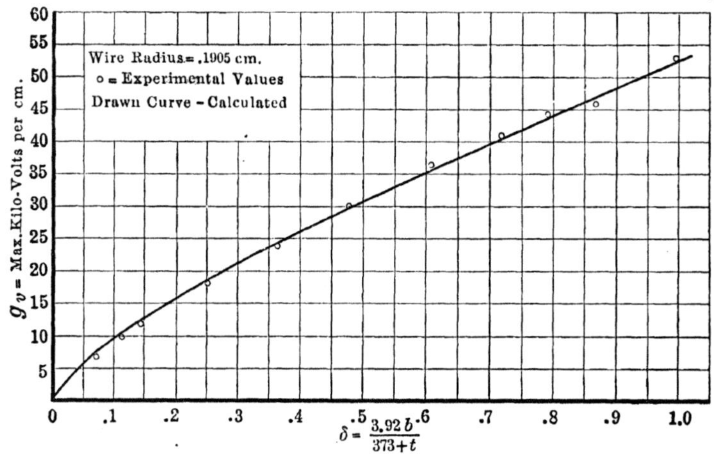

FIG. 37.—Effect of pressure on the strength of air.

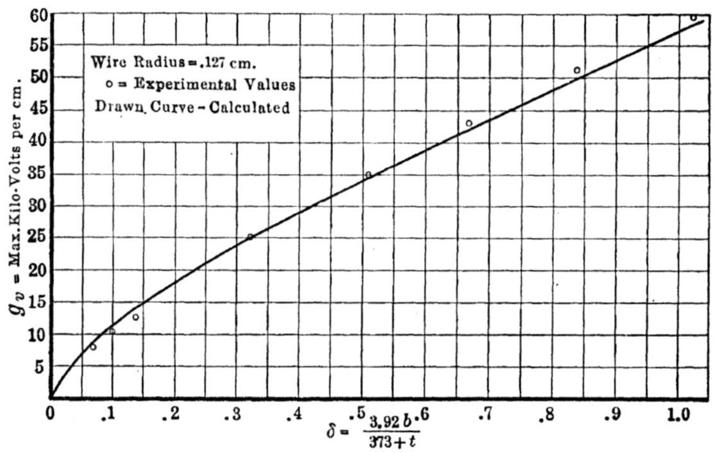

FIG. 38.—Effect of pressure on the strength of air.

a pressure range of 1.7 cm. to 76 cm. Table VIII is typical of observed and calculated values. Figs. 36, 37 and 38 show how these follow the previously predicted curve. A $\Sigma\Delta$ reduction of the values also confirms the above formulæ. For concentric cylinders

$$g_v = 31\delta\left(1 + \frac{0.308}{\sqrt{\delta r}}\right) \text{ max. kv. per cm.}$$

For parallel wires

$$g_v = 30\delta\left(1 + \frac{0.301}{\sqrt{\delta r}}\right) \text{ max. kv. per cm.}$$

The change in g_v with the density is apparently due to change in the molecular spacing, and thus depends only upon the relative density and not the absolute density. Heavy CO_2 at the same pressure and temperature has the same strength as air. This was not checked on hydrogen.

Electric Strength of Air Films.—If a definite amount of energy is necessary to start rupture at a finite distance a from the con-

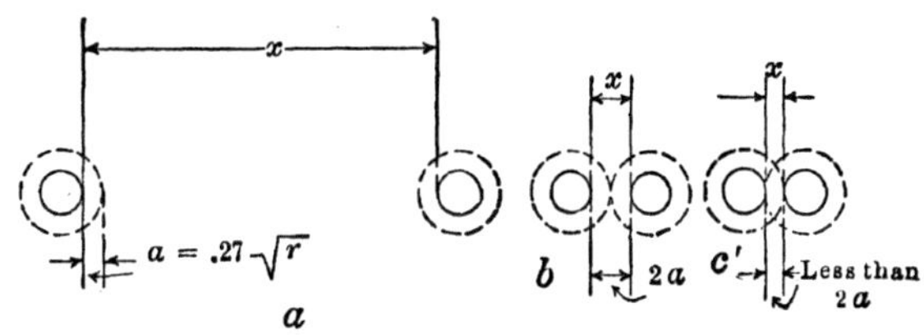

FIG. 39.

ductor, with a surface gradient g_v, it is interesting to speculate what will happen at very small spacings or when the distance between conductor surfaces is in the order of a. (See Fig. 39.)[1]

As the free "energy storage," "accelerating" or "ionizing" distance is then limited, a greater force or gradient should be required when the distance between the conductors approaches a. Experiments were made to determine this, using spheres as electrodes. The ideal electrodes for this purpose would be concentric cylinders, but the use of these as well as parallel wires at small spacing seemed impracticable. Spark-over and corona curves were made on spheres ranging in diameter from 0.3 cm. to 50 cm. and spacings from 0.0025 cm. to 50 cm. This discussion applies only to spacings up to $2R$ where corona cannot form.

[1] Actually, the "energy-zone" is not as shown, as the field is distorted at these small spacings.

In these tests a 60-cycle sine wave voltage was used. For the small spacing the spheres were placed in a very rigid stand. One shank was threaded with a fine thread, the other was non-adjustable (Fig. 40). In making a setting the adjustable shank was screwed in until the sphere surfaces just touched, as indicated by completing the circuit of an electric bell and single cell of a dry battery. A pointer at the end of the shank was then locked in place, after which the shank was screwed out any given number of turns or fraction of turns as indicated on the stationary dial. For larger spacings other stands were used.

A typical spark-over spacing curve and corona spacing curve is shown in Fig. 41. Theoretically, up to a spacing of $2R$ corona cannot form but spark-over must be the first evidence of stress.

Practically, corona cannot be detected at 60~ until a spacing of $8R$ is reached. This is because up to this point the difference

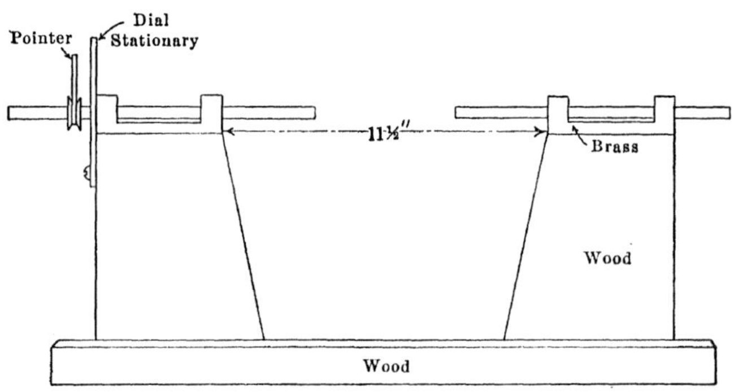

FIG. 40.

between the corona starting points and the spark point is very small. Above $8R$ the spark-over curve approaches a straight line as in the case of the needle gap curve. The corona curve above $2R$ and the spark curve below $2R$ are apparently continuous. The gradient curve, Fig. 42, is calculated from the voltage curve, Fig. 41. Where the spacing is less than $0.54\sqrt{R}$ the gradient increases first slowly and then very rapidly with decreasing spacing. Between $X = 0.54\sqrt{R}$ and $2R$ the gradient is very nearly constant. Above about $3R$ spacing the gradient apparently increases. This apparent increase is probably due to the effect of the shanks, etc., which become greater as X is increased. The effect of the shanks is to better distribute the flux on the sphere surface, and cannot be taken account of in the equation for gradient. This was shown experimentally by using different sizes of shanks at the larger spacing. Thus, when the spacing is greater than $3R$ the sphere is not suitable for studying the strength of air, as the gradient cannot be conveniently calculated. It is hence not a suitable electrode for studying corona, as corona does not form until the spacing is greater than $2R$. The maximum gradient at the surface of a sphere (non-grounded) may be calculated from the equation.

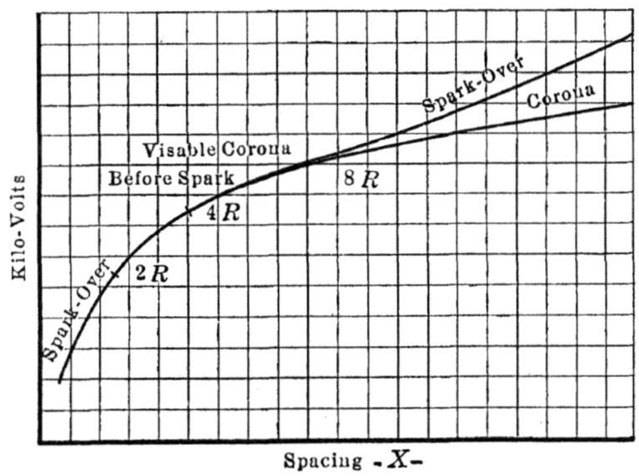

FIG. 41.—Variation of corona and spark-over voltages with spacing for spheres.

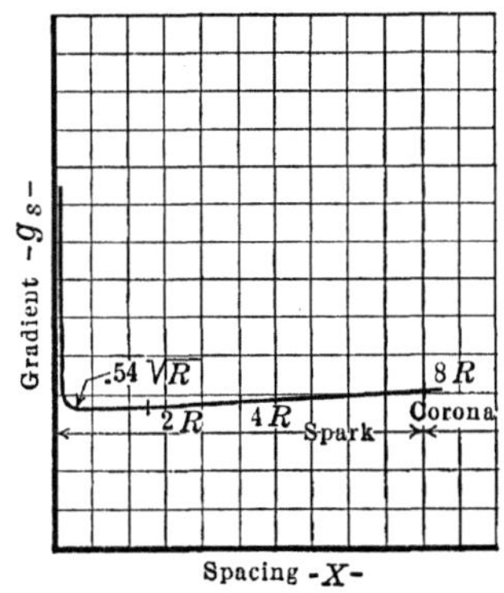

FIG. 42.—Variation of strength of air with spacing for spheres.

$$g = \frac{E}{X} f \tag{13a}$$

where X is the spacing
E the voltage

$$f = 1/4\left(\frac{X}{R} + 1 + \sqrt{\left(\frac{X}{R} + 1\right)^2 + 8}\right)$$[1]

The gradient for the non-grounded case may be conveniently calculated by use of the table on page 27. The gradient on the line connecting the sphere centers at any distance a from the sphere surface may be calculated from the complicated equation[2]

$$g_a = \frac{E}{X}\left\{\frac{2X^2\left[X^2(f+1) + 4\left(\frac{X}{2} - a\right)^2(f-1)\right]}{\left[X^2(f+1) - 4\left(\frac{X}{2} - a\right)^2(f-1)^2\right]}\right\} \tag{26}$$

Some of the experimental values are given in Tables IX and X. Typical voltage gradient curves are shown in Figs. 43, 44 and 45.

[1] G. R. Dean, G. E. Review, March, 1913.
[2] G. R. Dean, also Physical Review, Dec., 1912, April, 1913.

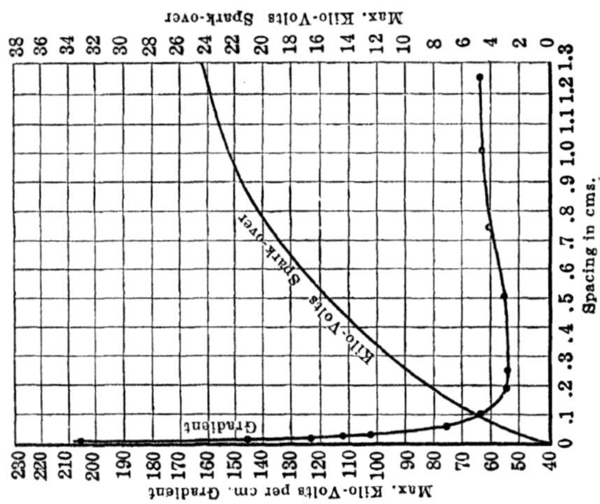

Fig. 43.

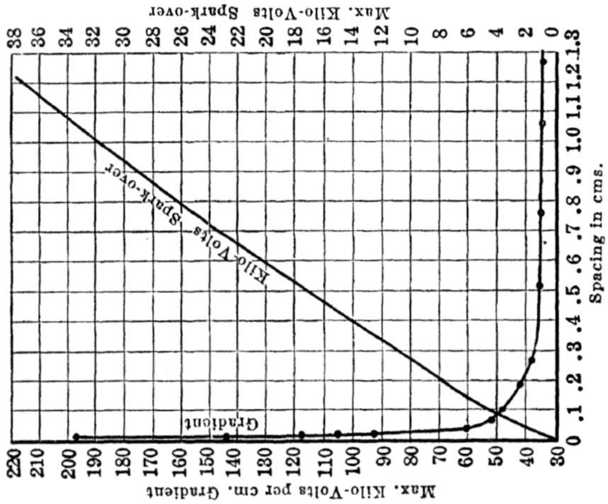

Fig. 44.

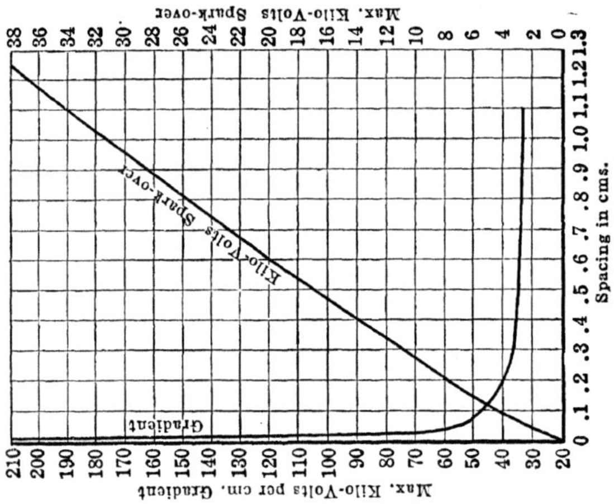

Fig. 45.

FIGS. 43, 44 and 45.—Variation of strength of air and spark-over voltages with spacing. Fig. 43.—Radius of sphere, R = 0.238 cm. Fig. 44.—R = 3.33 cm. Fig. 45.—R = 6.25 cm.

TABLE IX.—SPARK-OVER OF SPHERES AT SMALL SPACINGS
Brass Spheres R = 3.33 cm., Diameter = 2-5/8 in.

X Spacing In.	Cm.	$e_{eff.}$ Kv.	$\delta = \frac{3.92b}{273+t}$	$e_{max.} = \frac{\sqrt{2}(e_{eff.})}{\delta}$ (corrected)	$g_{max.} = \frac{e_{max.}}{X} f$ kv./cm.	X/R
0.001	0.00254	0.363	1.028	0.497	196.0	0.00076
0.002	0.00508	0.531	1.030	0.729	143.6	0.00152
0.003	0.00762	0.654	1.027	0.899	118.1	0.00228
0.004	0.01016	0.775	1.026	1.07	105.2	0.00305
0.005	0.0127	0.845	1.016	1.17	92.3	0.00382
0.010	0.0254	1.07	1.000	1.52	60.0	0.00764
0.020	0.0508	1.86	1.002	2.62	51.8	0.01528
0.040	0.1016	3.27	1.002	4.62	45.9	0.03056
0.075	0.1905	5.43	1.002	7.66	41.0	0.05730
0.100	0.254	6.92	1.002	9.77	39.5	0.0764
0.200	0.508	12.40	1.002	17.50	36.3	0.1528
0.300	0.762	17.70	1.002	25.00	35.4	0.2292
0.400	1.016	22.70	1.002	32.00	34.8	0.3056
0.500	1.270	27.75	1.002	39.20	34.9	0.382

TABLE X.—SPARK-OVER OF SPHERES AT SMALL SPACINGS
Brass Spheres R = 12.5 cm., Diameter. = 9.84 in.

X Spacing In.	Cm.	$e_{eff.}$ Kv. (read)	$\delta = \frac{3.92b}{273+t}$	$e_{max.} = \frac{\sqrt{2}(e_{eff.})}{\delta}$ (corrected)	$g_{max.} = \frac{e_{max}}{X} f$ kv./cm.	X/R
0.005	0.0127	0.807	1.023	1.116	87.9	0.00101
0.010	0.0254	1.220	1.021	1.689	66.5	0.00203
0.020	0.0508	2.050	1.031	2.849	55.9	0.00406
0.040	0.1016	3.38	1.022	4.68	46.2	0.00813
0.100	0.254	7.03	1.020	9.75	38.6	0.0203
0.200	0.508	12.54	1.012	17.44	34.8	0.0406
0.300	0.762	17.91	1.016	24.92	33.5	0.0609
0.400	1.016	22.82	1.010	31.76	32.0	0.0812
0.500	1.27	27.63	1.010	38.67	31.5	0.1016
1.000	2.54	53.0	1.000	74.90	31.6	0.2032
1.500	3.81	75.3	1.000	106.30	30.9	0.3048
2.000	5.08	96.4	1.000	136.20	30.5	0.4064
2.500	6.35	117.4	1.000	166.00	30.7	0.5080
3.000	7.62	139.2	1.000	196.90	31.2	0.6096
3.500	8.89	158.0	1.000	223.40	31.3	0.7112
4.000	10.16	174.9	1.000	247.10	31.0	0.8128
4.500	11.43	190.8	1.000	269.50	31.0	0.9144
5.000	12.70	203.6	1.000	287.20	30.8	1.0160

In Table XI are tabulated, for different sizes of spheres, the spark-over gradient at the constant part of the curve, the average gradient between $X = 0.54\sqrt{R}$ and $X = 3R$, and the approximate minimum spacing at which the gradient begins to increase

TABLE XI.—MAXIMUM RUPTURING GRADIENTS FOR SPHERES

(Average for Constant Part of the Curve)

R Radius in cm.	Spacing X, where g_s begins to increase (cm.)	g_s max. kv./cm. for constant part of curve
0.159	0.18	63.8
0.238	0.25	55.6
0.356	0.26	51.4
0.555	0.40	46.9
1.270	0.51	40.0
2.540	0.85	36.8
3.120	....	35.8
3.330	0.95	34.8
6.25	1.30	32.5
12.50	2.0	31.3
25.00	....	30.0

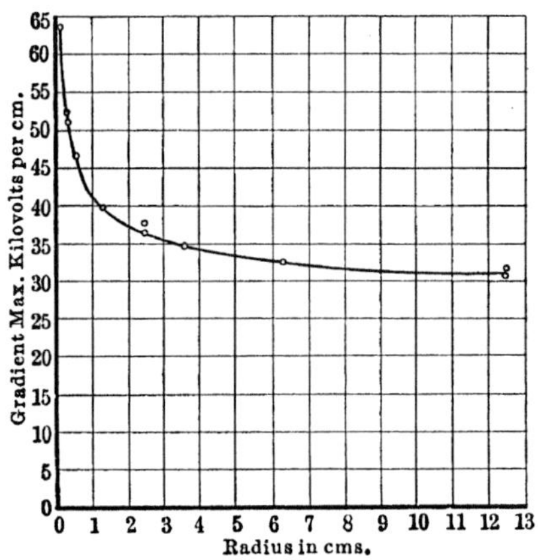

FIG. 46.—Variation of the apparent strength of air with sphere radius. Points measured gradient from constant part of curve. Drawn curve, calculated from equation (27).

in value. The gradient-radius curve is plotted in Fig. 46. This curve is very closely given by the equation

$$g_s = g_o\left(1 + \frac{\alpha}{\sqrt{R}}\right) \qquad (27)$$

which has exactly the same form as the similar curve for cylinders. The value of g_o is, however, lower than for the balanced field of a wire in a cylinder.

For a wire in a cylinder $g_o = 31 \quad g_v = 31\left(1 + \frac{0.308}{\sqrt{r}}\right)$ (25a)

For parallel wires $g_o = 30 \quad g_v = 30\left(1 + \frac{0.301}{\sqrt{r}}\right)$ (25b)

For spheres $g_o = 27.2 \quad g_v = 27.2\left(1 + \frac{0.54}{\sqrt{R}}\right)$ (28)

It is probable that the true strength of air is 31 kv. per centimeter as represented by the balanced field, it is apparently less for parallel wires due to the unbalanced field, and still less for spheres where the field is unbalanced to a greater extent.

The curve between the sphere radius and the approximate minimum spacing below which the gradient begins to increase appreciably is plotted in Fig. 47 from Table XI. The curve is represented by the equation

$$X = 0.54\sqrt{R} \qquad (29)$$

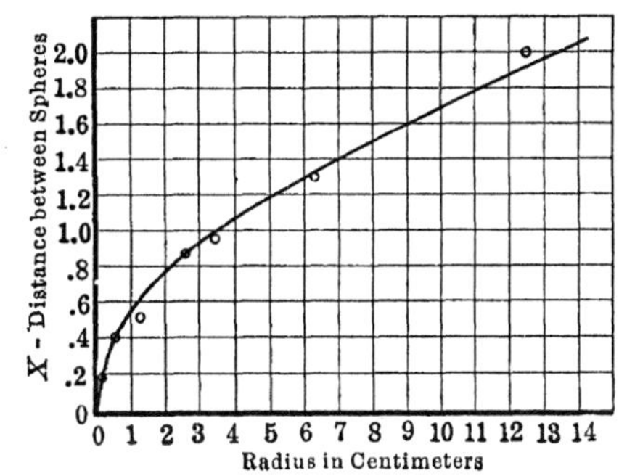

FIG. 47.—Sphere spacing below which apparent strength begins to increase.

which means that when the spacing is less than $0.54\sqrt{R}$ the gradient increases in value, at first slowly, then very rapidly.

It is now interesting to investigate the meaning of equation (27). In Fig. 48 the exact gradient is plotted from equation (26) for different distances from the sphere surface on the line connecting the centers and at given spacings as indicated by the small diagram in the upper corner of the figure. It is seen that for small distances from the sphere surface the curves for the different spacings fall together. Over the small range the gradient g_a at any point a centimeters from the sphere surface on the center line may be found approximately from

$$g_a \cong \frac{E}{(R + 2a)}\left(\frac{R}{X}\right)f \qquad (30)$$

The only reason for giving this approximation, which holds only for very small values of a and is only true when $a = 0$, is that (26) is too complicated to handle. The error due to this approximation is shown in Table XII.

TABLE XII

R = 1.27 cm. Energy distance, $a = 0.27\sqrt{R} = 0.3$ cm.				R = 12.5 cm. Energy distance, a = 0.95 cm.			
a	g_a, Exact	g_a, Approx.	X	a	g_a, Exact	g_a, Approx.	X
0.0	39.9	39.9	0.76	0.0	31.3	31.3	4
0.1	34.6	34.0	0.76	0.2	30.3	30.3	4
0.2	31.9	30.0	0.76	0.4	29.5	29.4	4
0.4	29.7	24.1	0.76	0.6	28.6	28.5	4
				0.8	28.3	27.7	4
0.0	39.9	39.9	1.21	1.0	27.5	26.9	4
0.1	34.2	33.4	1.21				
0.2	30.5	29.5	1.21	0.0	31.3	31.3	10
0.4	26.5	23.7	1.21	0.2	30.3	30.3	10
				0.4	29.5	29.4	10
				0.6	28.6	28.7	10
				0.8	28.0	27.7	10
				1.0	27.5	26.9	10

Then

$$g_s = \frac{(E_s)}{(X)} f = \frac{E_s}{R}\left(\frac{R}{X} f\right) \text{ exact mathematical.} \tag{13a}$$

$$g_s = g_o\left(1 + \frac{\propto}{\sqrt{R}}\right) \text{experimental for approximately constant part of curve.} \tag{27}$$

$$g_a \cong \frac{E}{(R + 2a)}\left(\frac{R}{X} f\right) \text{ approximate mathematical.} \tag{30}$$

Equating (13*a*) and (27)

$$\frac{E_s}{R}\left(\frac{R}{X} f\right) = g_o\left(1 + \frac{\propto}{\sqrt{R}}\right)$$

$$g_o = \frac{E_s}{(R + \propto\sqrt{R})}\left(\frac{R}{X}\right) f$$

which is the same form as (30) and means that at a distance

$$\frac{\propto\sqrt{R}}{2} = \frac{0.54\sqrt{R}}{2} = 0.27\sqrt{R} \text{ cm.}$$

from the sphere surface the gradient at rupture is always approximately constant and is g_o. As breakdown must take place at approximately $a = 0.27\sqrt{R}$ cm. from the sphere surface, the gradient should begin to increase at the spacing $2a = 0.54\sqrt{R}$. This is approximately so as shown in Fig. 47 and equation (30). The increase is at first slow at $X = 2a$, and very rapid at $X = a$. Fig. 48 shows that at $a = 0.27\sqrt{R}$ the gradient is not exactly constant for different spacings, or the curves do not fall together.

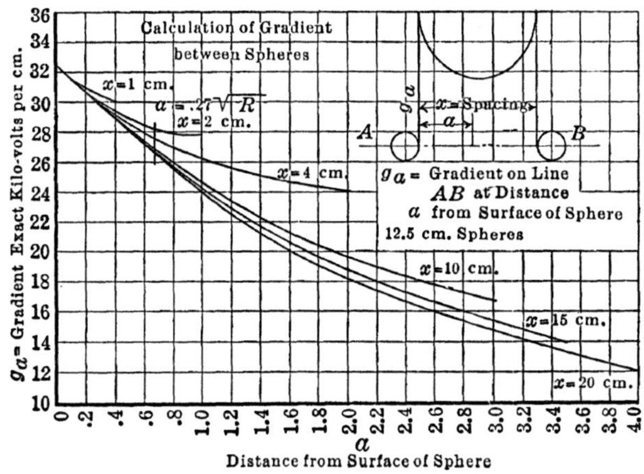

FIG. 48.—Gradient at different points on line through centers of spheres. Calculated from (26).

This means that g_o and $\propto$ in equation (27) cannot be exactly constant for a given radius, but must also be a function of X. This is experimentally shown to be the case, as there is a slight variation over the range $X = 0.54\sqrt{R}$ and $X = 2R$.

Influence of Frequency on the Visual Gradient.—The effect

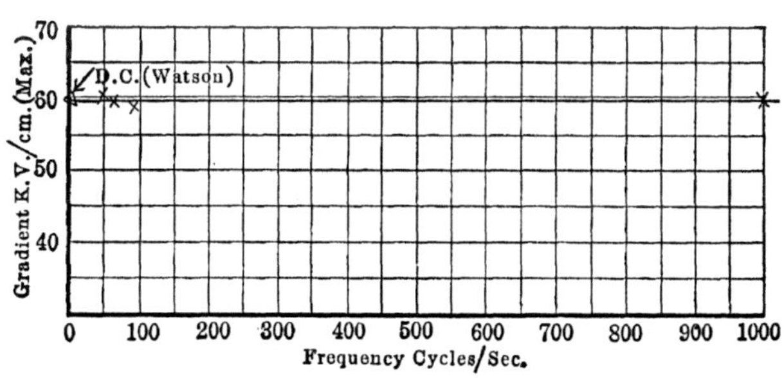

FIG. 49.—Variation of the apparent strength of air with frequency.

of frequency on g_v for the practical range of 25 to 60 cycles, if any, is very small and can be neglected. A few measurements are shown in Fig. 49. For the test range it is difficult to tell whether the slight variations are due to changes in wave shape or to frequency. There is a possibility of frequency entering this as a function:

$$g_v = g_o\left(1 + \frac{\alpha}{\phi(f)\sqrt{\delta r}}\right)$$

Investigations up to 1000 cycles show very little if any change. In this investigation the sine wave voltage was measured with a static voltmeter calibrated at 60 cycles. Measurement at 30,000 cycles (sine wave from a generator) made by the static voltmeter showed a slightly lower voltage than at 60 cycles.[1] Direct current points by Watson are given in Fig. 49. Over the commercial range of frequency, however, there is no appreciable effect of frequency.

Effect of Oil, Water or Dirt on the Visual Corona Point.—These tests were made in a manner exactly similar to the dry tests.

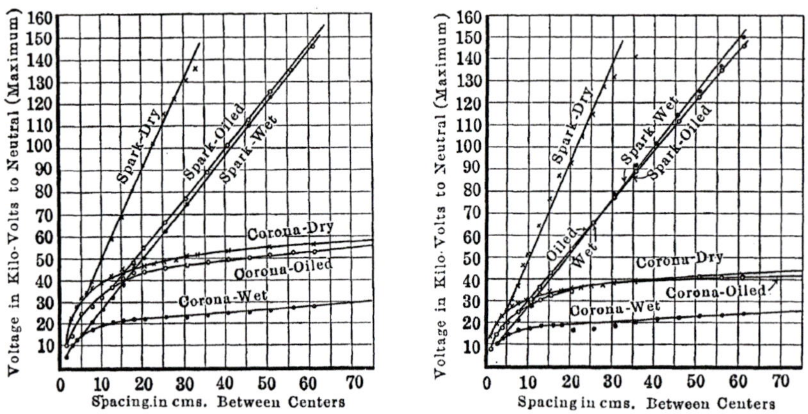

FIGS. 50 and 51.—Spark-over and corona voltages for parallel wires. (Wire surfaces dry, wet, and oiled. Max. kv to neutral. $s = 1$. Fig. 50.—Wire radius 0.205 cm. Fig. 51.—Wire radius 0.129 cm.)

In the oil tests, the surface of the wire was coated with a thin, even film by means of an oiled cloth. For the wet tests water was sprayed on the conductor surface before each reading by means of an atomizer. Figs. 50 and 51 are dry, wet and oil curves for two different sizes of wire.

For spark-over, both water and oil have approximately the same effect, that is, give very nearly the same spark-over voltage for all sizes of conductor. The curves very closely follow the needle gap curve.

For corona, water very greatly lowers g_v. Oil lowers g_v but to a

[1] See pages 106, 107.

much less extent than water. When the conductor is very small the per cent. increase in diameter due to oil more than compensates for the lowering effect. The *approximate* apparent visual corona gradient for oil and water coated conductors may be found

Water surfaces by fine spray or fog

$$g_v \cong 9\left(1 + \frac{0.815}{\sqrt{r}}\right) \text{ max. kv. per cm.}$$

Oil film surfaces

$$g_v \cong 19\left(1 + \frac{0.65}{\sqrt{r}}\right) \text{ max. kv. per cm.}$$

See Fig. 52.

If a water coated wire above the visual corona point is examined in the dark, it has the appearance of an illuminated atomizer. The surface quickly becomes dry.

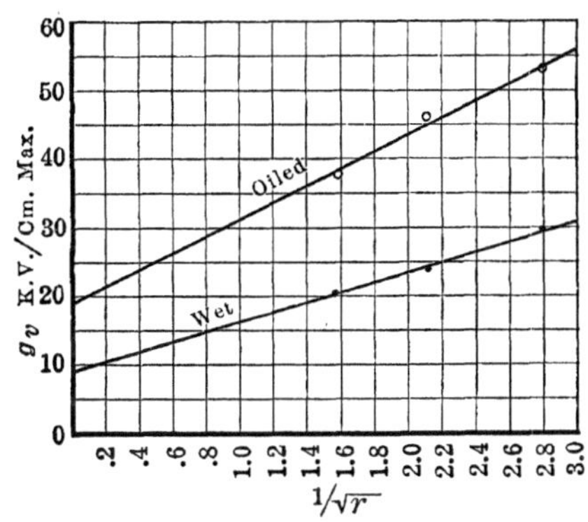

FIG. 52.—Apparent strength of air around wires with wet and oiled surfaces.

Dirt on the surface of the conductor, by increasing the gradient, causes local brush discharges, and if the surface is rough, corona starts at a lower voltage. This is taken care of in the formulæ by an irregularity factor m_v. Thus for a weathered or oxidized wire

$$g_v = g_o m_v \delta\left(1 + \frac{0.301}{\sqrt{\delta r}}\right)$$

The corona starting voltage for wet or oiled wires may be found by calculating g_v in above equations and substituting in equation (20).

Conductor Material.—With the same surface condition the visual corona point is independent of the material. This is shown in Table II.

Humidity.—Tests made over a very wide humidity range show that humidity has no appreciable affect upon the starting point of visual corona. After corona is present humidity has an effect on the spark-over voltage. This is discussed in Chapter IV.

Ionization.—Change of initial ionization of the air even to a considerable extent has no appreciable effect on the starting point of corona. This was found by test by increasing the voltage on the wire in the cylinder until considerably above the corona voltage, and then, while the cylinder was full of ionized air, lowering the voltage below the corona point and again raising it until glow appeared. The starting point was not appreciably changed. Initial ionization should, however, effect the time in which the discharge takes place. This will be discussed later.

Current in Wire.—A test was made to see if heavy currents flowing in the wire would change the starting point of visual corona. The test arrangement was as shown in the Fig. 53.

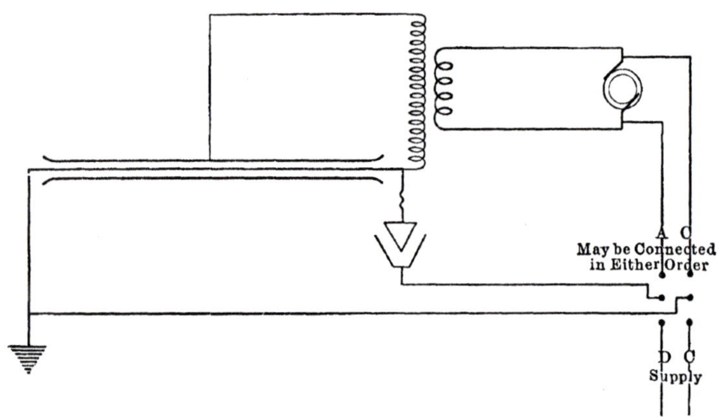

FIG. 53.—Apparatus for measuring corona starting point with current in the wires.

Tests were made with both a.c. in phase and out of phase, and d.c. in the wire. The results are given in Tables XIII, XIV and XV. The temperature of the wire was measured by the resistance method. The values in the last column are all corrected from the wire temperature to $\delta = 1$. If the current has no appreciable effect, as appears to be the case, these should all be equal. The variation is probably due to difficulty in getting the

TABLE XIII.—CORONA STARTING POINT WITH CURRENT FLOWING THROUGH WIRE

Concentric Cylinders

Radius of wire............................0.129 cm.
Radius of cylinder........................5.46 cm.

Kv. eff.	Amp. d.c.	Average temperature		Barom. cm.	δ (using wire temp.)	g_v	g_v reduced to $\delta = 1$
		Air	Wire				
20.0	0.0	26	26	75.6	0.991	58.4	58.9
19.7	15.2	26	32	....	0.970	57.5	59.2
19.4	31.4	26	48	....	0.922	56.7	61.4
18.5	39.0	26	64	....	0.878	54.1	61.6
17.8	49.2	26	82	....	0.824	52.0	63.1
16.9	75.0	26	120	....	0.753	49.4	65.6
20.4	0.0	23	23	....	1.000	59.6	59.6
20.3	18.0	23	27	....	0.987	59.3	59.9
19.4	33.0	23	50	....	0.917	56.7	61.8
17.9	50.8	24	91	....	0.813	52.2	64.2
12.9	99.6	24	320	....	0.500	37.6	75.2
16.4	67.6	26	102	....	0.790	47.9	60.6
13.9	78.6	26	212	....	0.610	40.8	67.0
16.1	61.6	26	144	....	0.710	47.0	66.2

TABLE XIV.—CORONA STARTING POINT WITH CURRENT FLOWING THROUGH WIRE

Concentric Cylinders

Radius of wire............................0.205 cm.
Radius of cylinder........................5.46 cm.

Kv. eff.	Amp.	Average temperature		Barom. cm.	δ (using wire temp.)	g_v	g_v reduced to $\delta = 1$
		Air	Wire				
25.3	14 (d.c.)	23	23	76.2	1.01	53.15	52.6
25.2	0	23	23	76.2	1.01	53.0	52.5
25.2	14 (a.c. in phase)	23	23	75.8	1.005	52.8	52.6
25.1	14 (a.c. out of phase)	23	23	75.8	1.005	52.6	52.3
25.1	0	23	23	75.8	1.005	52.6	52.3
25.2	0	24	24	75.25	0.994	52.8	53.2
25.2	14 (d.c.)	24	30	75.25	0.974	52.8	54.3
25.2	34 (d.c.)	25	60	75.25	0.886	52.4	59.1
23.8	56 (d.c.)	29	60	75.25	0.886	50.0	56.3
23.3	70 (d.c.)	30	73	75.25	0.853	48.8	57.3
21.8	114 (d.c.)	.....	110	75.25	0.771	45.7	59.2
25.0	60 (d.c.)	31	31	75.25	0.971	52.4	53.9
25.0	0	31	31	75.25	0.971	52.4	53.9
24.8	23 (a.c.)	31	44	75.25	0.931	52.0	55.8

TABLE XV.—CORONA STARTING POINT WITH CURRENT FLOWING THROUGH WIRE

Concentric Cylinders

Radius of wire.................................. 0.476 cm.
Radius of cylinder.............................. 5.465 cm.

Kv. eff.	Amp. d.c.	Average temperature		Barom. cm.	δ (using wire temp.)	g_v	g_v reduced to $\delta = 1$
		Air	Wire				
36.5	0	27	27	75.5	0.985	44.3	45
36.5	80	27	27	75.5	0.985	44.3	45

exact temperature. The air immediately surrounding the wire is assumed to be at the same temperature as the wire. Current flowing in a wire thus does not appreciably effect the corona point unless the temperature of the wire is increased.

Stranded Conductors or Cables.—While the visual critical corona point is quite sharp and definite for wires, it is not so for

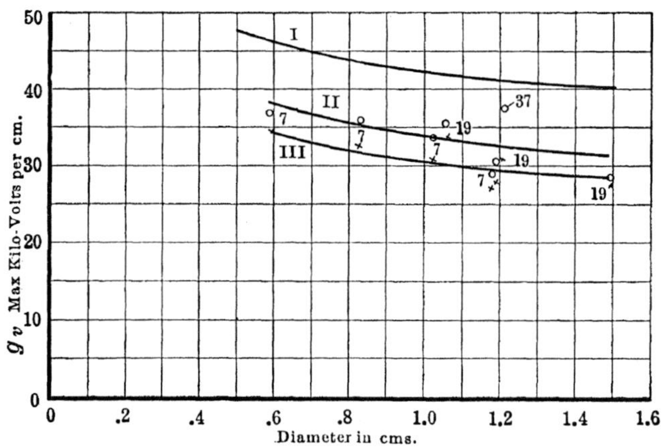

FIG. 54.—Apparent visual critical corona voltages for parallel cables. Numerals denote number of strands: I, polished copper wire; II, decided corona on cable, o; III, local corona all along cable, x.

cables or standard conductors. Corona, after it first appears, increases gradually for a considerable range of voltage until a certain definite voltage is reached where the increase is very rapid. The first point has been called the local corona point, and the second point the decided corona point. The curve, Fig. 54, for these corona points is compared with the curve for a smooth conductor. The starting point for cables may be found by the use of an irregularity factor, m_v.

$$g_v = g_o m_v \left(1 + \frac{0.301}{\sqrt{r}}\right) \text{kv. per cm.}$$

where m_v = .82 for decided corona
m_v = .72 for local corona
r = overall radius of cable.

This applies to cables of six strands or over.

It is interesting to note that for the decided corona point the visual critical voltage of a cable is about 3 per cent. lower than that of a wire with the same cross-section, or, more exactly "the diameter of a solid wire with the same critical voltage is about 97 per cent. that of the wire having the same cross-section as the cable."[1] This is shown in Table XVI.

TABLE XVI.—EFFECT OF STRANDING
Whitehead, A.I.E.E, June, 1911, Table III

Cables, strands outer layer	Diameter over all (*A*)	Diameter solid of equal section (*B*)	Diameter solid of equal critical volts (*C*)	*C/B*	*C/A*	Pitch of spiral	
						Cm.	Diameters
3	0.349	0.272	0.247	0.907	0.708	3.81	10.9
4	0.404	0.332	0.320	0.965	0.792	3.49	8.6
5	0.45	0.381	0.370	0.971	0.822	4.44	9.9
6	0.49	0.430	0.420	0.975	0.857	6.02	12.3
7	0.541	0.480	0.465	0.969	0.868	6.66	12.3
8	0.589	0.530	0.516	0.975	0.877	6.35	10.8
9	0.64	0.581	0.567	0.977	0.886	6.98	10.9
......	......	....	....	.....	.....	....	....
3	0.336	0.27	0.307	0.767	0.616	None	None
4	0.378	0.312	0.25	0.802	0.665	None	None

Conductors of the Same Potential Close Together.—When conductors of the same potential are arranged close together the critical breakdown voltage is much greater than that of a single conductor or when they are far apart. The simplest case, that of two, is shown in Fig. 55. The two conductors for a given test were kept at a constant distance $S/2$ from the ground plate. Potential was applied between the conductors and plate. The separation m was then varied and critical voltages read at differ-

[1] Whitehead, Dielectric Strength of Air, A.I.E.E., June, 1911.

ent spacings. Refer to Fig. 56 (0.163-cm. wire 30 cm. from neutral plane). When $m = 0$, $e_v = 31.5$ eff. As the spacing m was increased, e_v increased to a maximum of 35.8 kv. With increasing m, e_v then gradually decreases to a constant value which is the same as that for a single wire. The maximum voltage is about 5

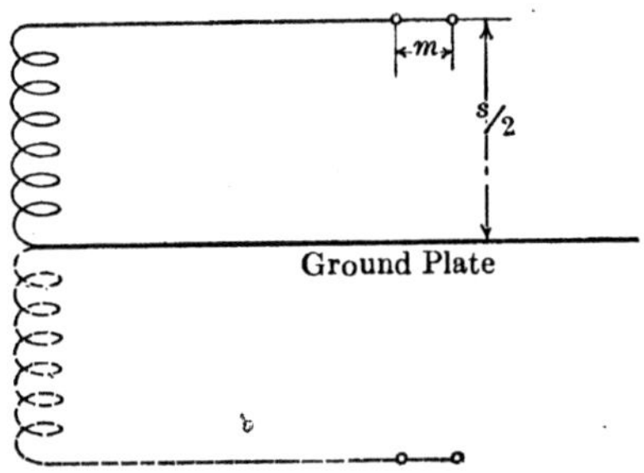

FIG. 55.—Arrangement for two conductors at same potential, and plate.

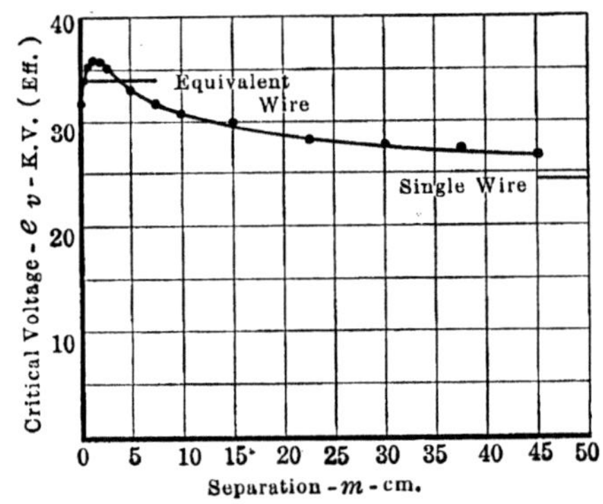

FIG. 56.—Critical voltage on two conductors at the same potential and various separations (see Fig. 55).

per cent. greater than the critical voltage of a single conductor of the same cross section.

With the same amount of conductor material, much higher voltages can be used without corona loss when the conductor is

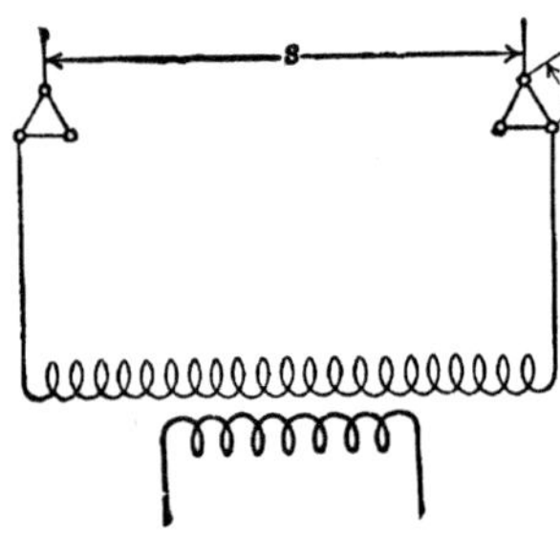

FIG. 57.—Conductors of the same potential arranged in a triangle.

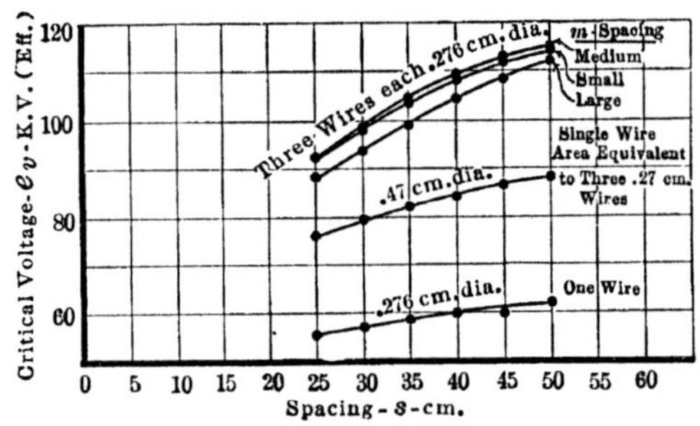

FIG. 58.—Critical voltages for conductors arranged as in Fig. 57.

split up into three or more small conductors, properly arranged, than with a single conductor. The results of tests made on a single-phase line with split conductors arranged in a triangle as in Fig. 57 are given in Fig. 58. Fig. 58 shows curves for a single split wire and also for a single wire of a cross section equal to that of the three split conductors. Fig. 59 shows how the voltage varies with varying *m*.

With the split conductor arrangement, in the special case given, the critical voltage is from 20 to 30 per cent. greater than that of a single wire containing the same amount of material.

Whitehead has made similar tests on three wires in a triangle and also four wires placed on a square in the center of a cylinder.

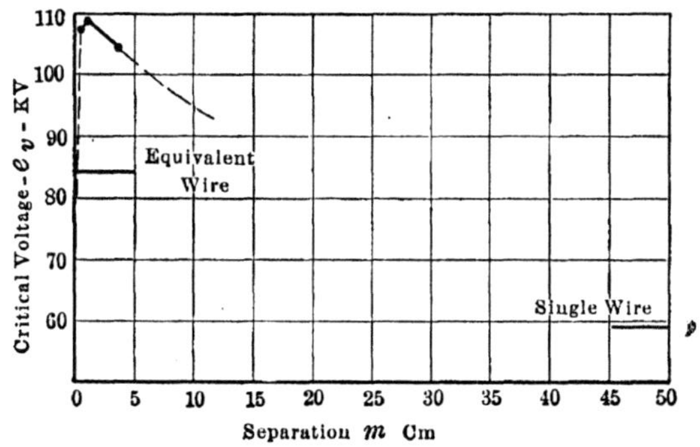

FIG. 59.—Critical voltage of conductors arranged as in Fig. 57. (S = constant = 40 cm., *m* varying.)

He finds 16 per cent. increase for three wires and 20 per cent. for four over the voltage of a single conductor of the same cross section.

STROBOSCOPIC AND PHOTOGRAPHIC STUDY

Photographic Study.—A photographic study of corona on wires and cables was made as follows: Two parallel conductors were spaced 122 cm. between centers. The camera was focused on one conductor only. The distance to the lens was such as to show the conductors at approximately actual size. An exposure was made for a given time at a given voltage. The plate was then shifted slightly, the voltage increased and a second exposure was made for the same time. That is, a given series shows the same part of the same single wire at different voltages. This operation was repeated until the series for a given wire was complete. A glass lens was used unless otherwise stated. (See Fig. 60.) These photographs are shown in Figs. 61 to 68.

Photographs 67 and 68 were made to show the effects of moisture. In Fig. 67 the stranded cable was brought up to the critical point. Water was then thrown on the cable. The result is shown in Fig. 68. What was a glow at the surface of the dry cable became at the wet spots, a discharge extending from 5 to 8 cm. from the conductor surface. The discharge has the appearance of an illuminated atomizer.

Diameter of Corona.—On a smooth wire the boundary line of corona appears to be fairly definite. The apparent visual diameter may be measured by viewing through a slit. The apparent diameter may also be found photographically. If the photograph is made through a quartz lens the ultraviolet rays will not be cut off from the plates as when a glass lens is used.

Whitehead has made some measurements on the **apparent** diameter, comparing the visual method and the photographic method with both quartz and glass lenses. He finds that the apparent diameters are respectively by the visual, glass lens, and quartz lens methods in the ratios of 1 : 1.6 : 1.9.[1] It therefore appears that there is a considerable content of the corona at the ultraviolet which is not visible to the eye. As soon as corona appears it seems to have a definite finite thickness.

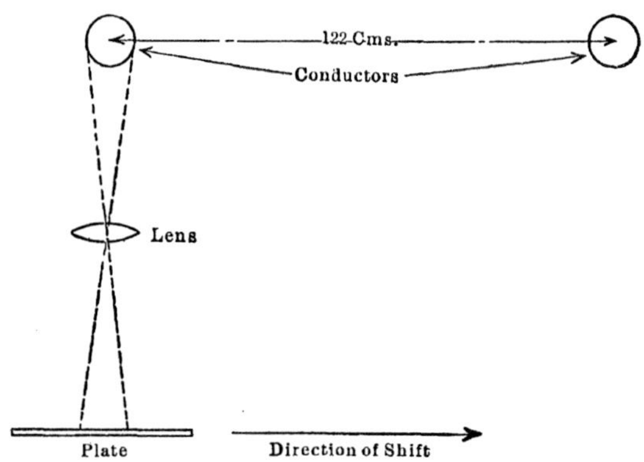

FIG. 60.—Method of making corona photographs.

TABLE XVII.—DIAMETER OF CORONA ON WIRE IN THE CENTER OF A CYLINDER

Diameter wire 0.233 cm., 18.6-cm. cylinder
Fig. 9 (Whitehead, A.I.E.E., June, 1912)

Kilovolts	Diameter corona, mm.	Time, min.	Lens	No.	Diameter corona, visual method, mm.
22.5	5.5	2	Glass	(a)	
27.5	9.3	2	Glass	(b)	
32.5	11.1	2	Glass	(c)	6.7

[1] Whitehead, Electric Strength of Air, A.I.E.E., June, 1912.

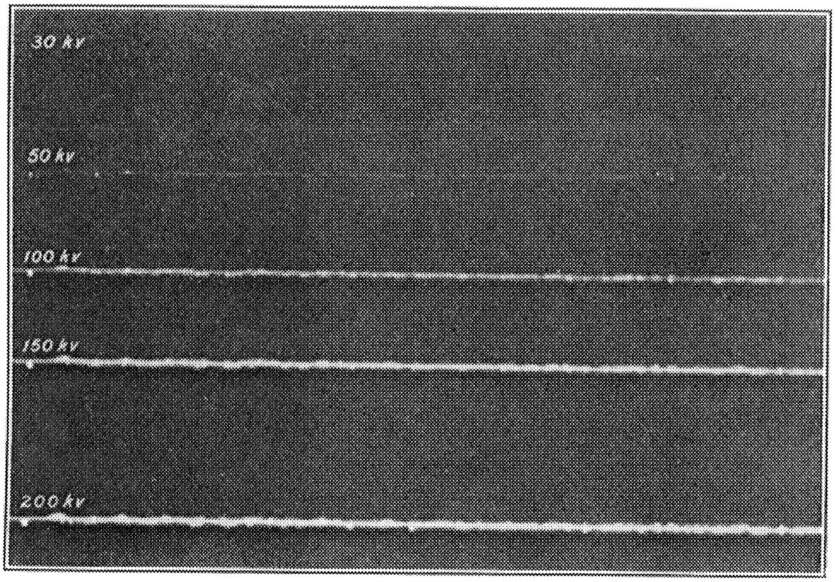

Fig. 61.—Corona on bright tinned phosphor-bronze wire. Diameter, 0.051 cm.

Fig. 62.—Corona on copper wire polished after each exposure. Diameter, 0.186 cm.

(*Facing page* 74.)

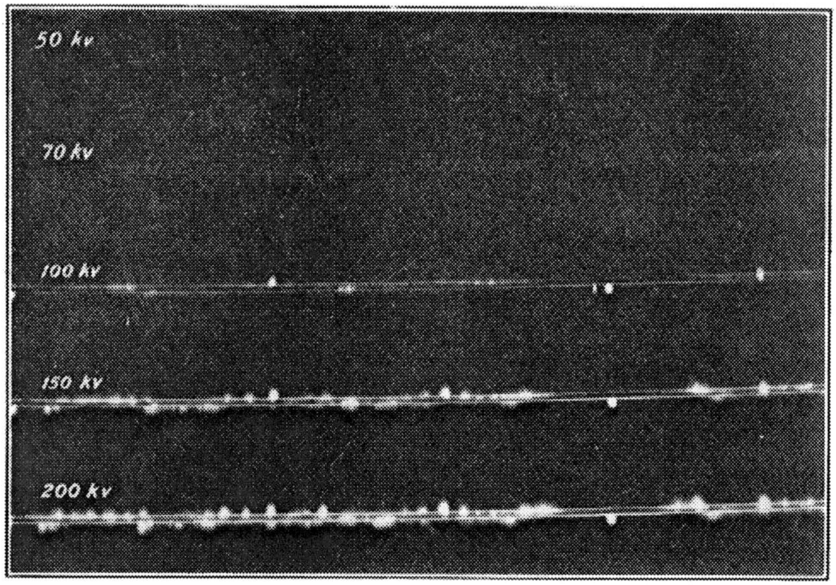

FIG. 63.—Corona on a polished copper wire. Diameter, 0.186 cm. Operated at 200 kv., then allowed to stand idle. (This shows effect of oxidation.)

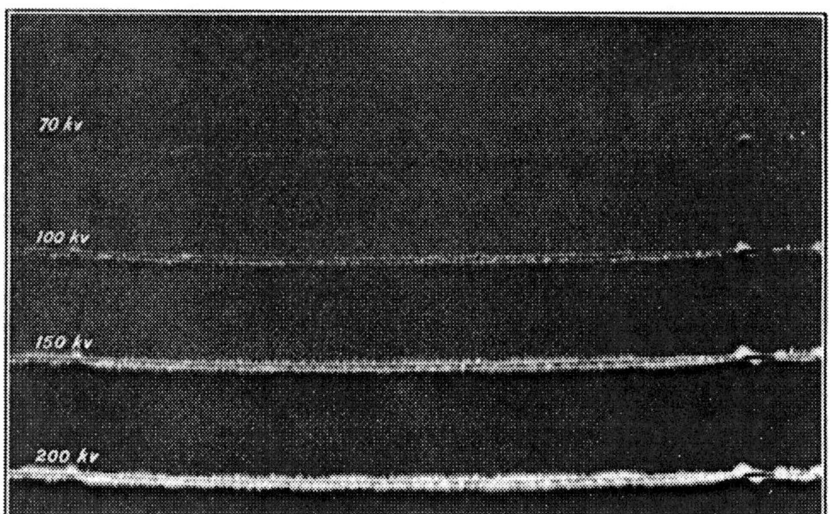

FIG. 64.—Corona on a weathered galvanized iron wire. Diameter, 168 cm.

FIG. 65.—Corona on a 1.25 cm. polished brass rod and unpolished copper cable.

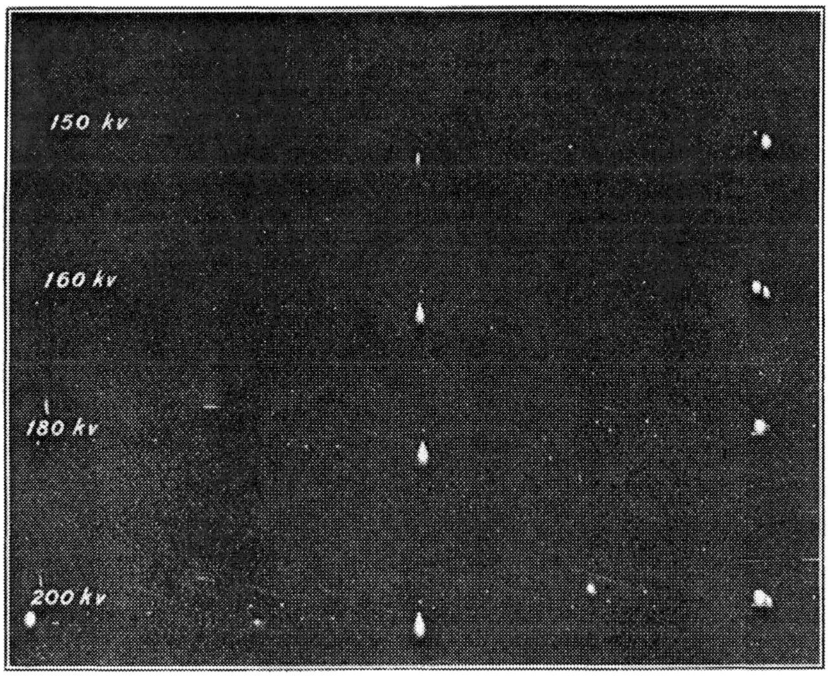

FIG. 66.—Corona on a No. 3/0 weathered cable.

(*Following Fig.* 64.)

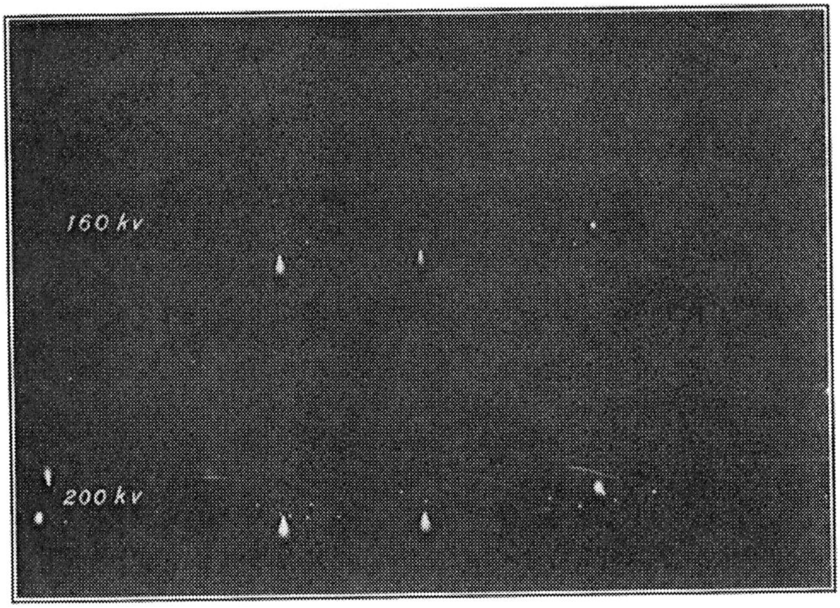

FIG. 67.—Corona on a No. 3/0 line cable. Dry.

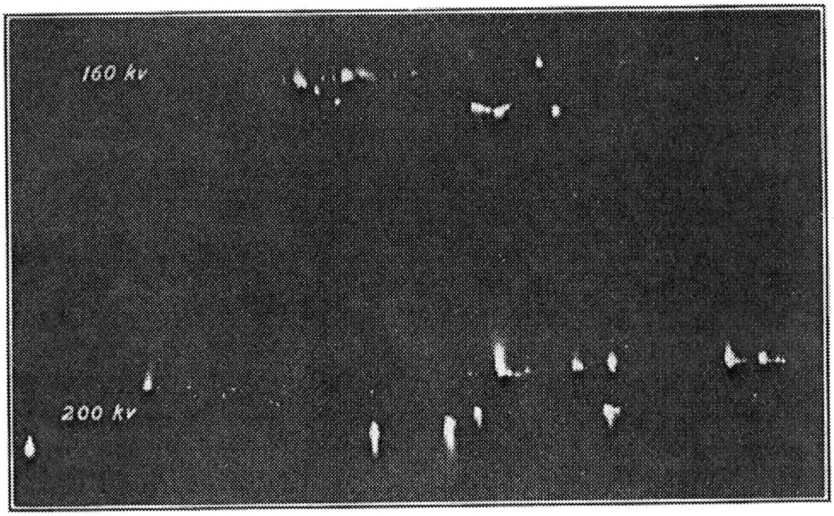

FIG. 68.—Corona on a No. 3/0 line cable. Wet.

(*Facing page* 75.)

TABLE XVIII.—DIAMETER OF CORONA, WITH AND WITHOUT ULTRAVIOLET CONTENT

(On a Wire in the Center of a Cylinder)

Exposure 20 min., 32.5-kv. cylinder, diam. = 18.6 cm.

(Whitehead, A.I.E.E., June, 1912)

0.232-cm. wire		0.316-cm. wire		0.399-cm. wire	
Quartz and glass	Quartz alone	Quartz and glass	Quartz alone	Quartz and glass	Quartz alone
12.0	12.4	12.5	13.3	11.4	13.0
11.5	12.6	13.0	14.0	11.6	13.0
11.5	12.6	13.0	14.0	12.0	12.8
11.4	13.0	12.8	14.0	12.0	12.8
11.6	12.6	12.8	13.8	11.7	12.9

Figs. 69, 70 and 71 and Tables XVII and XVIII taken from Whitehead are self explanatory. Fig. 72[1] shows the apparent diameter of corona on a given wire. At the start the corona

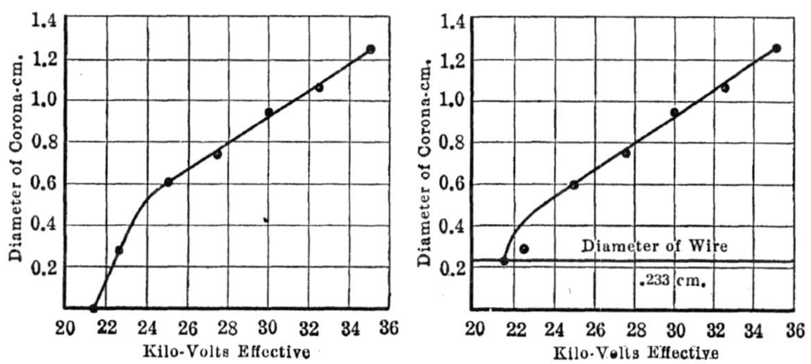

FIG. 72. FIG. 73.

Figs. 72 and 73.—Diameter of corona (0.233 cm. wire in 18.6 cm. cylinder).

appears to take immediately a definite finite thickness; the rate of increase is then quite rapid, but gradually assumes a linear relation.

[1] In Fig. 73 is a curve through the same points. Throughout this curve the correction of 1.18 has been used to include the ultraviolet. However, near the starting voltage the corona seems to be very largely ultraviolet. This explains the low point at 22.5 kv.

A study of the power loss equation leads one to suspect that the mechanism of corona loss is more complicated than might at first be supposed. This is also indicated by many peculiar phenomena of the spark discharge. For instance, while investigating a.c. spark-over and corona for parallel wires it was observed that when the end shields were not used, and the wires came directly in contact with the wooden wheel supports, corona often appeared to bridge completely between the conductors without a dynamic arc. In this case it seemed possible that the corona on the positive wire extended out farther than the corona on the negative wire and that the positive discharges overlapped and combine in the eye, giving the effect of a single discharge completely across between the conductors.

In the hope of throwing further light on the discharge and loss mechanism, an investigation of corona and spark was made with the help of the stroboscope.

A needle gap was first arranged across the transformer with a high steadying resistance. The impressed voltage was adjusted until corona appeared all the way between the conductors as in Fig. 74.

Examination of this was then made through the stroboscope which was so set that the right needle, Fig. 74(2), was seen when positive, and the left when negative. To the eye, the discharge from the positive needle has a bluish-white color and extends out a considerable distance, the negative appears as a red and hot point. The photograph shows more of the negative than is seen by the eye. Fig. 74(1) is the discharge as it appears without stroboscope, 74(2) with the right needle as positive, 74(3) with stroboscope shifted 180 deg. to show left needle as positive. In 74(4) the stroboscope has the same position as 74(3), but the voltage is higher, and many fine "static" sparks can be seen.

If voltage above the visual corona point is impressed on two parallel polished wires a more or less even glow appears around the wires. After a time the wires have a beaded appearance. On closer examination the beads appear as reddish tufts, while in between them appears a fine bluish-white needle-like fringe. On examination through the stroboscope it can be seen that the more or less evenly spaced beads are on the negative wire, while the positive wire has the appearance, if not roughened by points, of a smooth bluish-white glow. At "points" the positive discharge extends out at a great distance in the form of needles; it is possible that it always extends out but is not always visible except as sur-

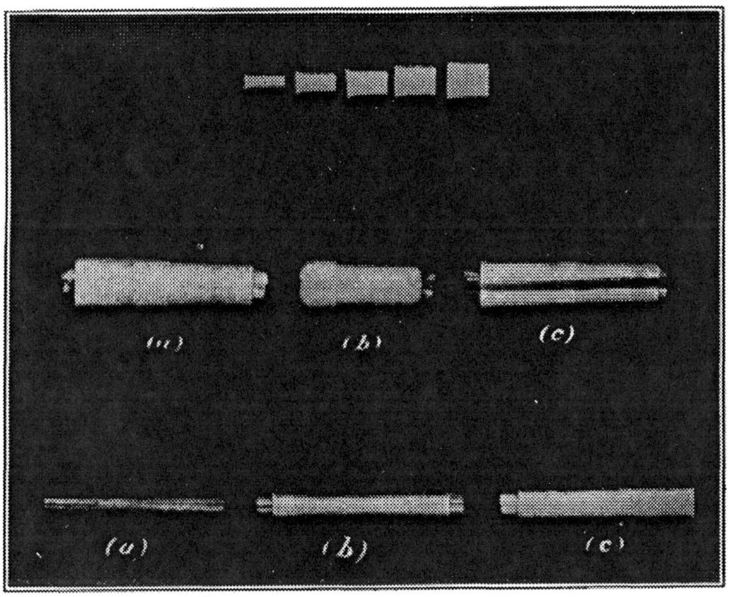

DIAMETER OF CORONA.

FIG. 69 (Upper).—Diameter of corona on 0.233 cm. wire in 18.6 cm. cylinder. $e_v = 21.5$ kv. Glass lens kilovolts 22.5, 25, 27.5, 30, 32.5 respectively. (Whitehead.)

FIG. 70 (Middle).—Diameter of corona showing effect of ultra-violet. (*a*), 0.232 cm. (*b*), 0.316 cm. (*c*), 0.399 cm. Left side of (*a*)(*b*)(*c*), quartz lens. Right side of (*a*)(*b*)(*c*), glass lens. (Whitehead.)

FIG. 71 (Lower).—Corona on 0.233 cm. wire. at 22.5, 27.5, 32.5 kv. $e_v = 20.75$ kv. Glass lens. (Whitehead.)

(*Facing page* 76.)

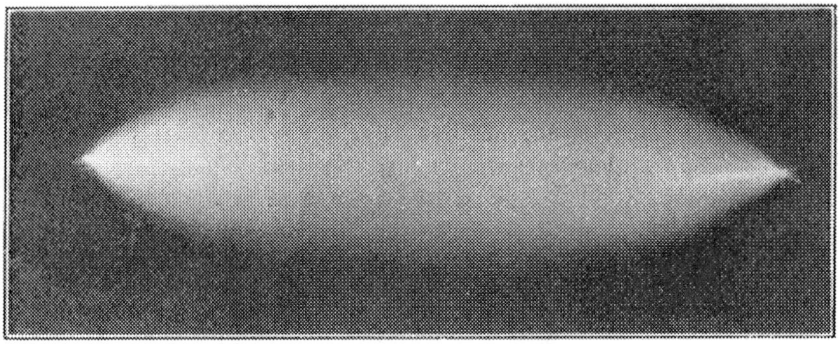

(1) Without stroboscope, 72,000 volts.

Left (−) (2) With stroboscope, 72,000 volts. Right (+)

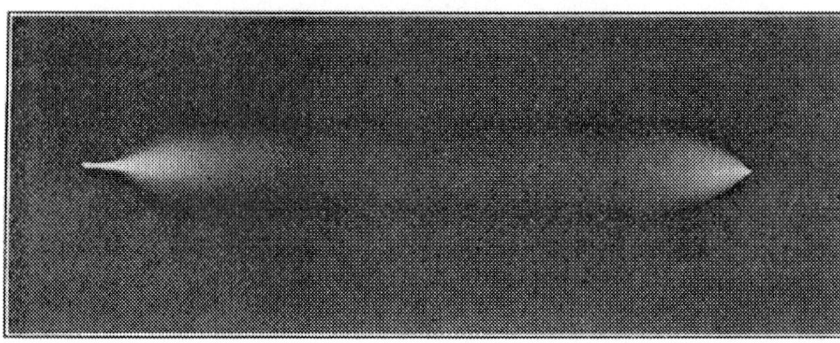

Left (+) (3) Same as (2), stroboscope rotated 180°. Right (−)

Left (+) (4) Same as (3), voltage increased to 84,000. Right (−)

Fig. 74.—Corona between copper needle points. 20.5 cm. gap.

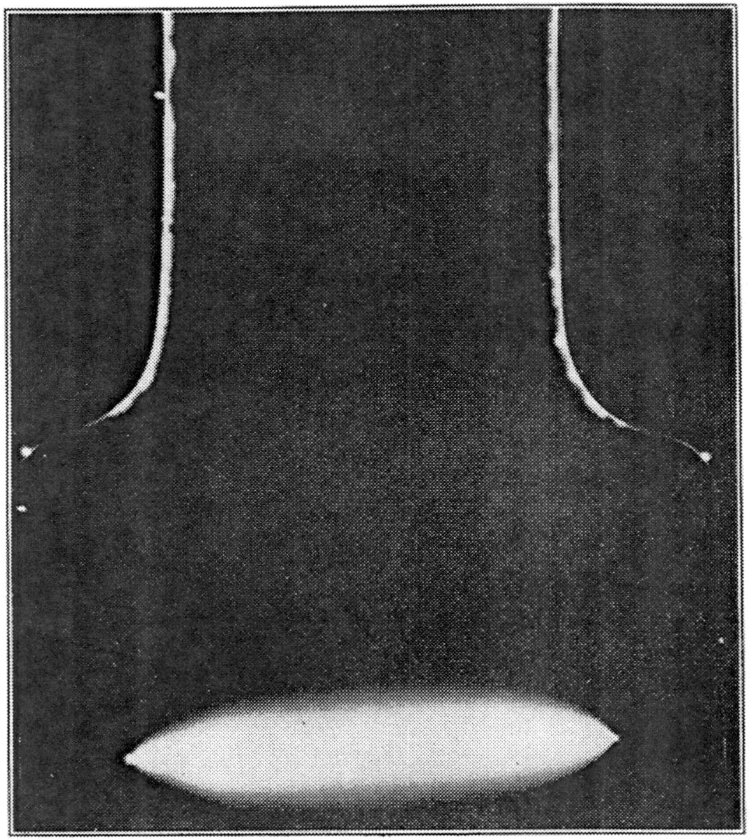

Fig. 75.—Comparison of corona on wires and needles. Phosphor-bronze wire, 14.5 cm. spacing, needles 18 cm. spacing. Volts, 71,000. Without stroboscope.

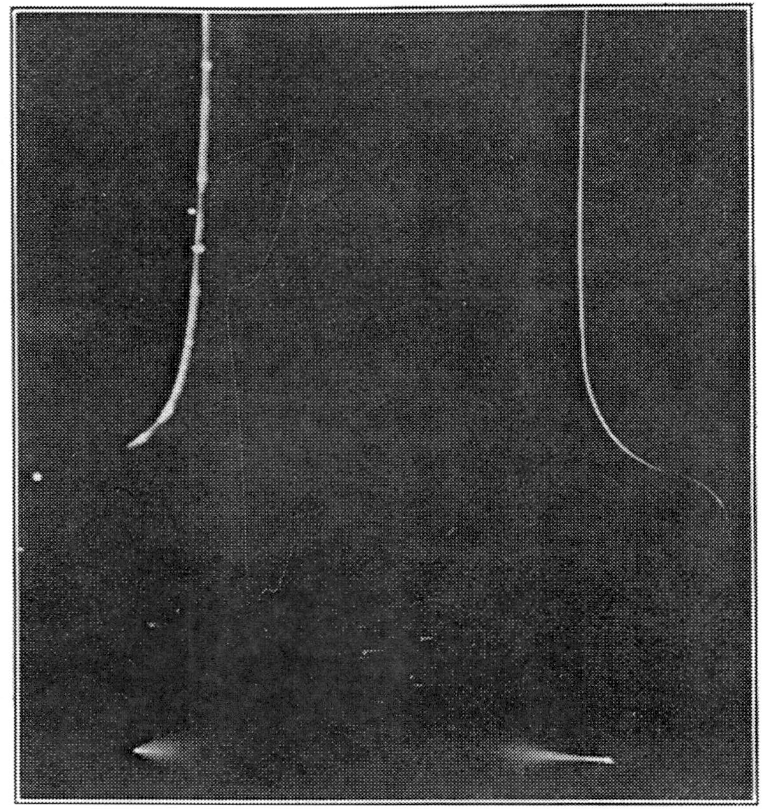

\- +

Fig. 76.—Comparison of − and + A. C. corona on wire and points. (Same as Fig. 75 with stroboscope.)

(Following Fig. 74.)

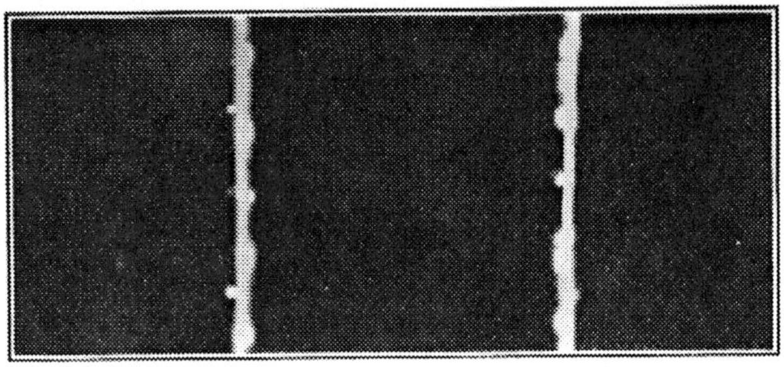

(1) Without stroboscope.

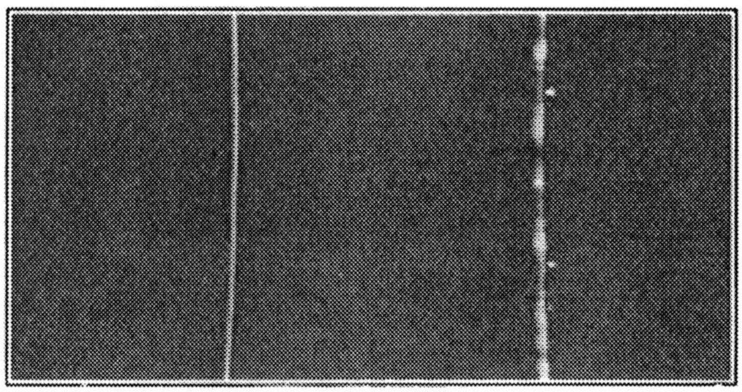

Left (+) (2) With stroboscope. Right (−)

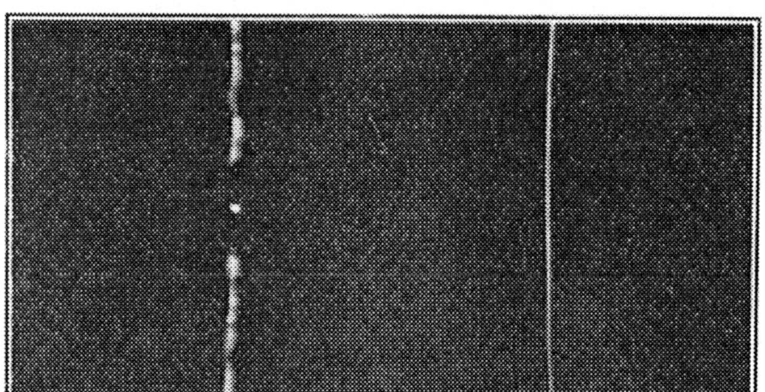

Left (−) (3) With stroboscope rotated 180°. Right (+)

Fig. 77.—Corona on parallel wires. No. 13 B. and S. copper wire. Spacing, 12.7 cm. Volts, 82,000.

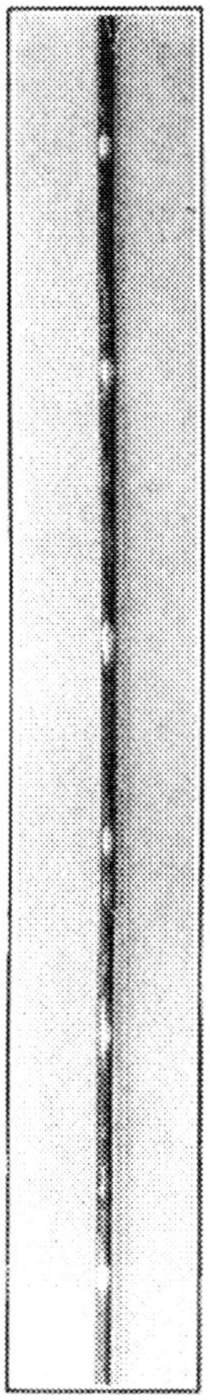

FIG. 78.—Section of wire (Fig. 77). "Dead." Bright spots position of negative "beads."

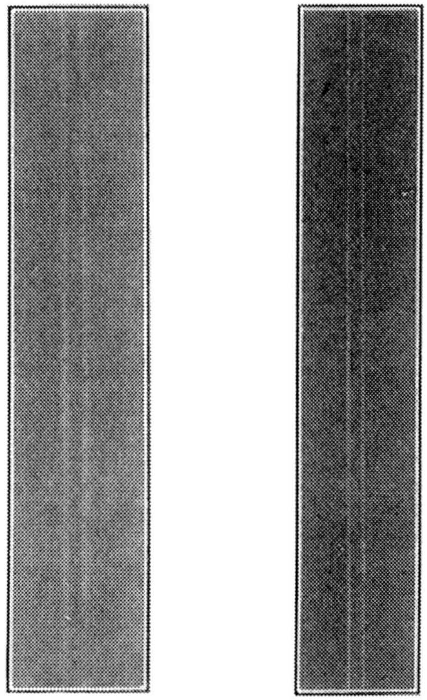

Left (−) Right (+)

FIG. 79.—Polished brass rod. Diameter, 0.475 cm. Spacing, 120 cm., Volts, 150,000. Note that negative "beads" are just starting to form.

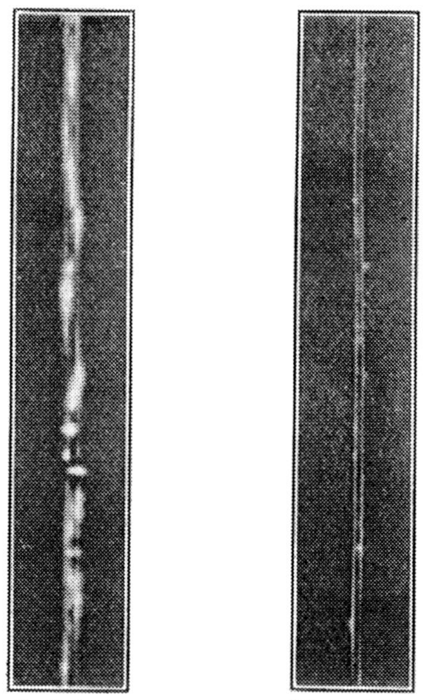

Left (−) Right (+)

FIG. 80.—Copper wire. Diameter 0.26 cm. Spacing, 120 cm. Volts, 200,000. Polished at start. Note negative corona apparently following spiral "grain" of wire.

(Following Fig. 77.)

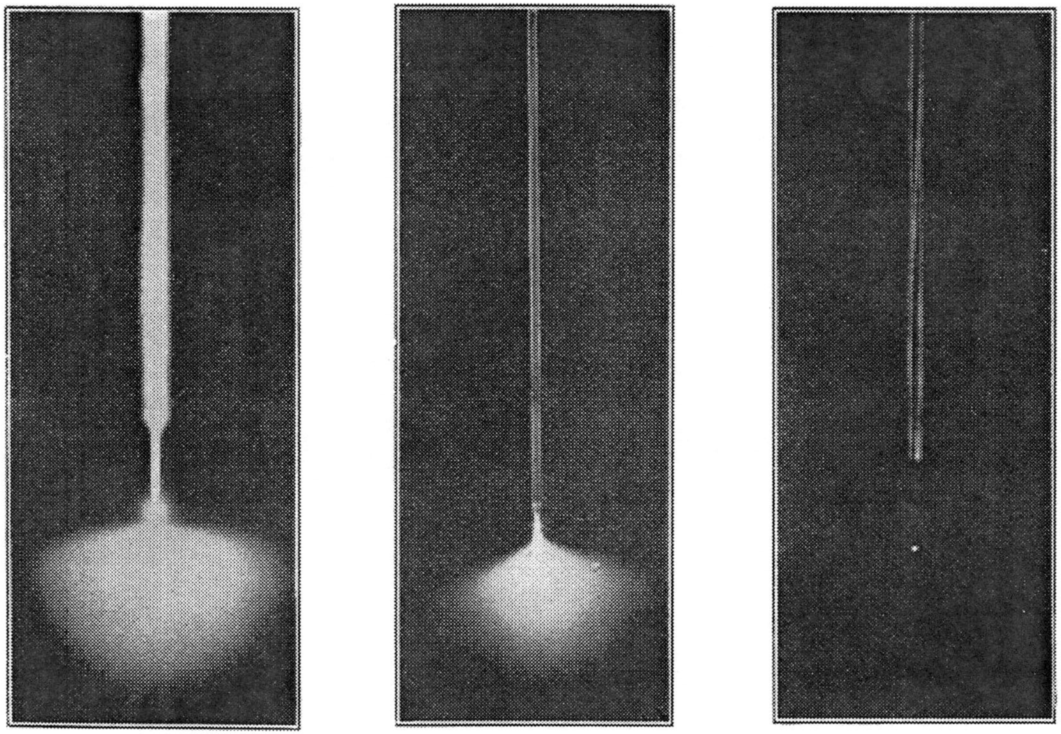

(1) Without stroboscope ±. (2) With stroboscope +. (3) With stroboscope −.

Fig. 81.—One of two parallel polished steel rods pointed at ends. Diameter, 0.325 cm. Spacing, 120 cm. Volts, 180,000.

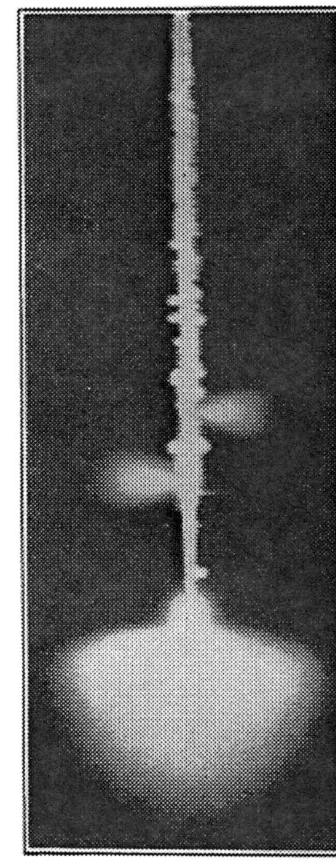

Fig. 82.—Same as Fig. 81 (1) Surface dirty.

(*Facing page* 77.)

face glow. Thus, the appearance of beads and fringe to the unaided eye is really a combination of positive and negative corona. In Figs. 75 and 76 two wires are placed close together at the top. The bottom is bent out and needles are fastened on. Fig. 75 is without a stroboscope. Fig. 76 is taken with a stroboscope set to show positive right and negative left. Thus, positive and negative coronas for points and wires are directly compared. Fig. 77(1) is taken without the stroboscope, (2) with right negative, (3) with stroboscope shifted 180 electrical degrees to show the right positive. These wires were, at the start, highly polished. At first corona appeared quite uniform, but, after a time, under voltage, the reddish negative tufts separated, more or less evenly spaced as shown. 24(2) is the same with stroboscope shifted 180 deg. Fig. 78 shows a section without voltage. The bright spots are still polished and correspond in position to the negative tufts. The space in between is oxidized. Thus, the negative discharge appears to throw metal or oxide from the surface at discharge points. This takes place with either copper or iron wire.

Fig. 79 shows positive and negative corona on wires widely spaced to get uniform field distribution. A close examination of the negative shows beads about to form. Fig. 80 shows a similar pair of conductors. The negative in this case has formed a spiral, apparently following the grain twist of the conductor.

A large fan-like bluish discharge is often observed extending several inches from the ends of transformer bushings, points on wires, etc. This discharge has the appearance of a bluish spray, reddish at the point. The stroboscope shows that the bluish spray is positive, while the red point at the base of the spray is negative. Fig. 81 shows one of two parallel polished rods (120 cm. spacing), supported at the top and brought to sharp points at the bottom; 81(1) shows how each wire appears without stroboscope; 81(2) is the wire when positive, 81(3) the wire when negative. Note the dark space on 81(3) between the point and negative corona spiral of tufts. 81(1) shows this space to have only the positive glow.

Water was placed on a pair of parallel conductors. At the wet places the positive corona extended out in long fine bluish-white streamers. (See Fig. 82 without stroboscope.) With certain forms of dirt on the wires the negative corona appears as red spots, the positive always as streamers. It is also interesting to note that if a uniformly rough wire is taken, as a galvanized wire or "weathered" wire, the positive appears as bluish needles, while the reddish negative is more uniform than on the "corona-spotted" polished wire, in which case the negative corona appears as concentrated at the non-oxidized spots. It is probable that the polished spots are kept so by metal and oxide being "thrown out" at the negative, as suggested above.

Mechanical Vibration of Conductors and Other Phenomena.—Several years ago a pair of 20 mil steel conductors, 500 ft. long, were strung at about 10-ft. spacing, for power loss measurements. It was noticed at high voltage that the conductors vibrated, starting with a hardly perceptible movement, which in a few minutes had an amplitude of several feet at the center of the span. Generally one wire vibrated as fundamental, the other as third harmonic. The period of the fundamental in this case was about one per second.

Figs. 83 and 84 show this condition repeated in the laboratory on short lengths of conductor. In Fig. 83 one wire is vibrating as the fundamental, the other as the second harmonic. The motion is rotary. For the wire with a node in the center, Fig. 83, it is extremely interesting to note that for about one-half of the rotation the wire appears very bright, for the other half rotation the wire is much less bright. This seems to mean that each part of the wire is rotating at the power supply frequency—60 cycles per second. Hence, it has the effect of the stroboscope, and for part of the rotation there is always negative corona and for the other part always positive corona.

Oscillograms of Corona Current.—Bennett has made some very interesting oscillograms of corona current.[1] Fig. 85(*a*), (*b*) and (*c*) shows the voltage wave applied between a cylinder and a concentric wire, and the resulting current. The part of the wave above the zero line occurs when the wire is $-$, and that below when the wire is $+$; (*a*) is for a voltage very slightly above the critical voltage and shows a very sudden sharp hump in the current wave when the wire is $+$, and a spread out hump when the wire is $-$. This gives the appearance of corona starting at a slightly lower voltage on the $+$ wire; (*b*) and (*c*) show the positive and negative humps at higher voltage. The oscillation is caused by the sudden "corona spark" discharging through the reactance and capacity of the circuit.

Some tests made on the starting time appear to show that several cycles are necessary for stable conditions.

[1] Bennett, An Oscillographic Study of Corona, A.I.E.E., June, 1913.

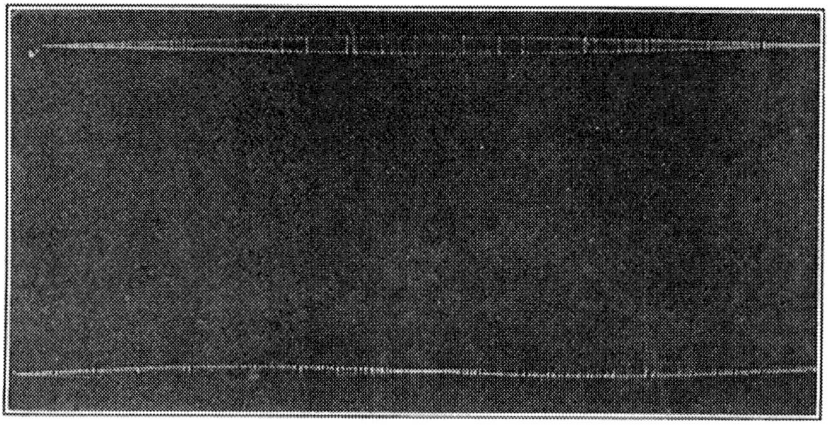

FIG. 83.—Mechanical vibration of parallel wires due to corona.

FIG. 84.—Mechanical vibration of parallel wires due to corona.

(*Facing page* 78.)

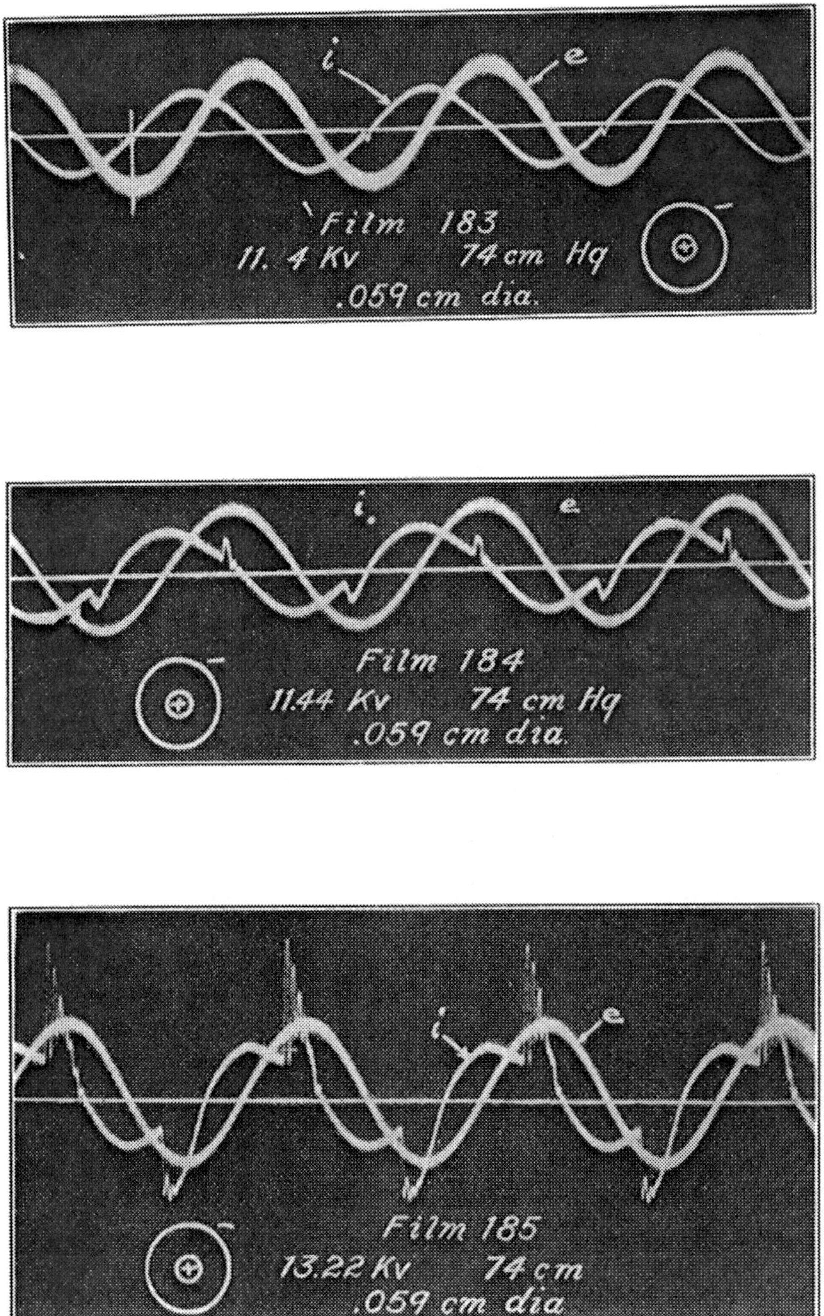

FIG. 85.—Oscilligrams of corona current. (Bennet, A. I. E. E., June, 1913.)

CHAPTER IV

SPARK-OVER

By spark-over is generally meant a disruption of the dielectric from one conductor to another conductor. Corona is the same phenomena—spark from a conductor to space or local spark-over.

Parallel Wires.—If impressed voltage is gradually increased on two parallel wires placed a considerable distance apart in air, so that the ratio S/r is above a certain critical value, the first evidence of stress in the air is visual corona. If the voltage is still further increased the wires become brighter and the corona has the appearance of extending farther out from the surface. Finally, when the voltage has been sufficiently increased, at some chance place a spark will bridge between the conductors. When the spacing is small, so that S/r has a critical ratio, spark and corona may occur simultaneously, or the spark may bridge across before corona appears. If the spacing is still further reduced so that S/r is below the critical ratio the first evidence of stress is complete spark-over and corona never appears. (See page 27.)

Extensive tests have been made.[1] The method of making tests was to start at the smaller spacings with a given value of r and measure the spark-over voltage. When the spacings were above the critical ratio of S/r, and corona formed before spark-over, the corona voltage was noted first. The voltage was then increased until spark-over occurred. The spark-over point is not as constant or consistent as the corona point and is susceptible to change with the slightest dirt spot on the conductor surface, and any unsteady condition in the circuit, humidity, etc. At the beginning of the tests it was found necessary, in order to get consistent results, to put water tube resistances in series with the conductors to eliminate resonance phenomena. These resistances were high, but not sufficiently so to cause an appreciable drop in voltage before arc-over.

Table XIX is a typical data table. Each point is the average of a number of readings.

[1] See Law of Corona II, A.I.E.E., June, 1912.

TABLE XIX.—CORONA AND SPARK-OVER FOR PARALLEL WIRES
Temperature 17 deg. C., bar. 75.3 cm.
Wire No. 0, diameter 0.825 cm.

Test No. 166 values read			No. 0 wire, corrected to 25° C., 76 bar.			
Spacing	Effective kv. to neutral		Maximum values to neutral		Maximum	
Cm. S	Corona e_v	Spark e_s	Corona e_v	Spark e_s	Corona g_v	Spark g_s
2.54	None	15.8	....	21.9	....	41.4
3.81	None	22.5	....	31.2	....	42.5
5.08	None	27.3	....	37.9	....	43.2
6.35	None	31.05	....	43.2	....	43.8
7.62	None	35.0	....	48.5	....	44.9
8.89	None	37.35	....	51.8	....	45.0
10.16	40.4	40.9	56.0	56.7	44.0	44.6
12.70	41.8	42.1	58.0	58.1	44.0	44.1
13.97	43.7	46.0	60.7	60.5	44.2	46.7
15.24	45.9	48.1	63.6	67.0	45.1	48.9
15.78	46.6	54.1	64.8	75.0	43.8	50.8
20.32	48.9	59.6	67.7	82.8	44.0	53.7
22.86	50.1	66.2	69.7	91.7	43.7	56.8
25.40	51.1	71.5	70.7	99.2	43.1	60.4
27.94	52.1	79.0	72.4	109.7	42.9	65.1
30.48	53.1	84.5	74.0	117.0	42.9	67.9
33.02	54.1	89.6	74.8	124.0	42.4	70.2
35.56	55.1	95.5	76.5	132.5	42.6	73.9
38.10	56.1	102.3	77.8	141.9	42.7	77.8
40.64	57.1	106.5	79.4	149.0	42.9	80.5
60.96	63.3	....	87.0	....	42.9	....

In columns 4 and 5 are voltages reduced to the maximum value to neutral and corrected to standard δ. Column 6 gives the surface gradient for corona. Column 7 gives the surface gradient for spark, up to the spacing where corona starts first; above this critical spacing it gives the apparent surface gradient as the conductor above this point must be larger on account of corona. As the field around the conductors at the small spacings is very much distorted it is necessary to use formula 12(a) or 12(b) to calculate the surface gradient.

Fig. 86 is a typical curve. Voltage is plotted with spacing for spark and corona. Up to spacing 12.4 cm. there is spark-over before corona. This curve seems to be continuous with the corona curve which starts at this point. The spark curve here branches and is very close to a straight line within the voltage range. In Fig. 87, the surface gradient curves are plotted. The

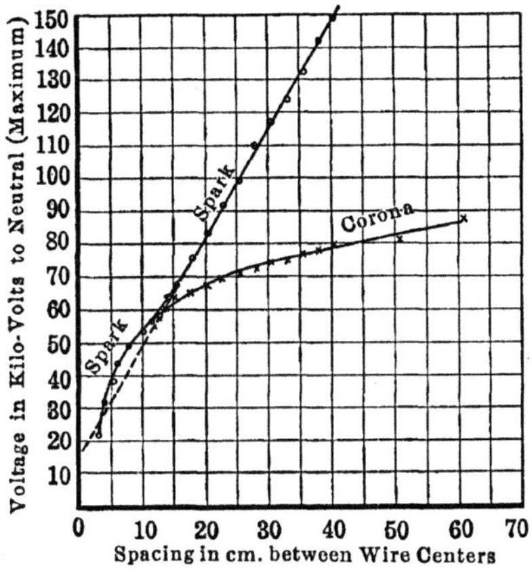

FIG. 86.—Spark-over and visual corona voltages.

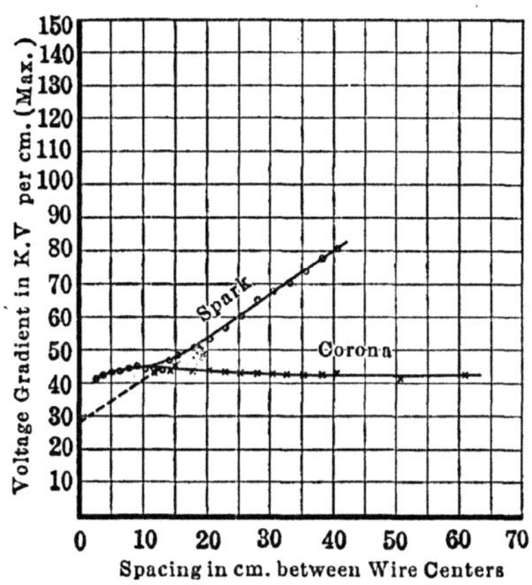

FIG. 87.—Corona gradient and apparent spark gradient.

(Parallel polished copper wires, 0.825 cm. diameter. $\delta = 1.$)

corona gradient is a straight line parallel to the X axis with a slight hump at the critical ratio of S/r. The *apparent* spark

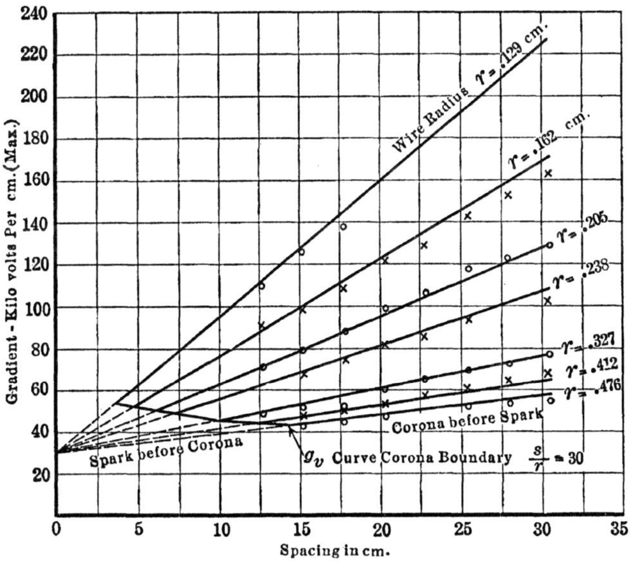

FIG. 88.—Apparent spark-over gradients for parallel wires. (Points measured, curves calculated.)

gradient is also a straight line, within the test range. It intersects the corona line at the critical ratio point, or at what may be termed the triangular point, and extended it cuts the g axis at $g = 30$. These are characteristic curves. (See also Figs. 50 and 51.) For a given spacing the spark-over voltage increases as the size of the conductor decreases.

It is important to note that for all sizes of wire the spark gradient curve extended as a straight line cuts the gradient axis at approximately $g = 30$. Spark curves extended as straight lines through the critical ratio point and intersecting the gradient axis at $g = 30$ are shown in Fig. 88. The triangular point or critical ratio of S/r is tabulated in Table XX. Its average value is

TABLE XX.—CRITICAL RATIOS S/r—EXPERIMENTAL VALUES

Intersection point of g_v and g_s

Size, B. & S.	Radius cond., cm.	S, cm.	S/r	Size, B. & S.	Radius cond., cm.	S, cm.	S/r
0	0.461	13.5	29.3	6	0.205	6.2	30.2
0	0.412	11.7	28.4	8	0.162	4.8	29.6
2	0.327	10.2	31.2	10	0.129	4.0	31.0
4	0.260	7.9	30.4	12	0.103	3.0	29.1
5	0.230	7.3	31.7			Average	30.1

$S/r = 30$. If it is assumed that the spark-gradient curve is a straight line the conditions are, that it must cut the corona gradient line at $S/r = 30$ and extended must cut the g axis at $g_o = 30$.

The gradient for g_v, or the gradient *at the* triangular point or *below* it, is

$$g_v = g_o\left(1 + \frac{0.301}{\sqrt{r}}\right) \tag{18}$$

therefore, the approximate apparent gradient *at or above* the triangular point is

$$g_s = g\left(1 + \frac{0.301}{\sqrt{r}}\frac{S}{r}\frac{1}{30}\right)$$

$$= 30\left(1 + \frac{0.01}{\sqrt{r}}\frac{S}{r}\right) \text{ kv. per cm. max.}$$

This follows because of the assumption of a straight line through two fixed points.

The approximate spark-over voltage above the triangular point is

$$e_s = g_s\, r \log R/r \text{ kv. to neutral max.}$$

Below the triangular point it may be found by substituting g_v for g_s.

In Fig. 88 each drawn curve is for g_s values calculated for varying spacing at constant radius. The points are measured values. The corona boundary line is the g_v curve; it intersects the g_s curves at $S/r = 30$. Corona does not form below this line, but spark jumps across immediately.

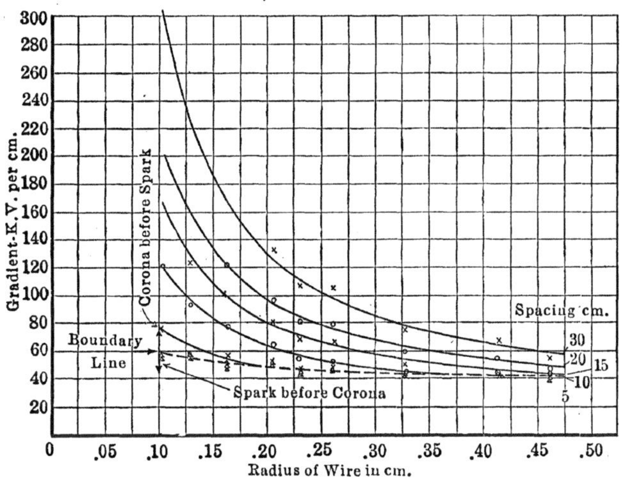

FIG. 89.—Apparent spark-over gradients for parallel wires. (Points measured, curves calculated.)

In Fig. 89 each curve is drawn for a constant spacing and varying radius. The broken line is the critical ratio line. It also corresponds to the g_v curve. For spacings below this line spark takes place immediately before corona forms, and the g_s values fall pretty well on the g_v line as shown by triangles.

Fig. 90 is voltage plotted in the same way. Below the corona boundary, where spark occurs before corona, the e_s curve does not hold. The broken lines are calculated from g_v and e_v. The points are observed values. Thus corona gradient and spark-over gradient, and hence spark voltage and corona voltage below $S/r = 30$, are the same.

No great accuracy is claimed for this formula. It may, however, be useful in approximately determining the arc-over between conductors in practice. Dirt or water, however, will greatly modify the results, as will appear below.

The reason that spark takes place before corona can form at small spacings or below $S/r = \propto$ is discussed on page 27 for concentric cylinders, in which case g was taken as constant.

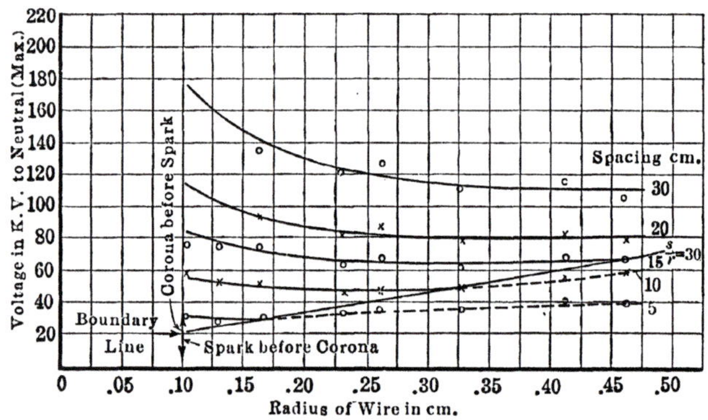

FIG. 90.—Spark-over voltages between parallel wires. (Maximum values to neutral given. Points measured, curves calculated.)

We know, however, that g_v is a function of r, and for air

$$g_v = g_o\left(1 + \frac{0.301}{\sqrt{r}}\right)$$

$$e = g_o\left(1 + \frac{0.301}{\sqrt{r}}\right) r \log_e R/r$$

Differentiating for maximum

$$\frac{de}{dr} = g_o\left(1 + \frac{0.301}{2\sqrt{r}}\right)\left(\log_e R/r - 1 - \frac{0.301}{\sqrt{r}}\right)$$

or, e is maximum when

$$\left(1 + \frac{0.301}{2\sqrt{r}}\right)\left(\log_e R/r - 1 - \frac{0.301}{\sqrt{r}}\right) = 0 \tag{31}$$

This gives a ratio of R/r greater than ϵ. The experimental ratio in Fig. 91 is 3 and checks with the above.

If a very small value of r is taken corona forms and then after the voltage is sufficiently increased, spark-over occurs. It might be supposed that with increasing voltage the center wire would become larger and larger in effect due to conducting corona and finally, when $\dfrac{R}{\text{radius} + \text{corona}}$ = critical ratio, spark-over would occur. This is not the case. It takes a much higher voltage for the small wire corona than for metallic cylinders with R/r at maximum ratio. Hence, corona seems to be either in *effect* a

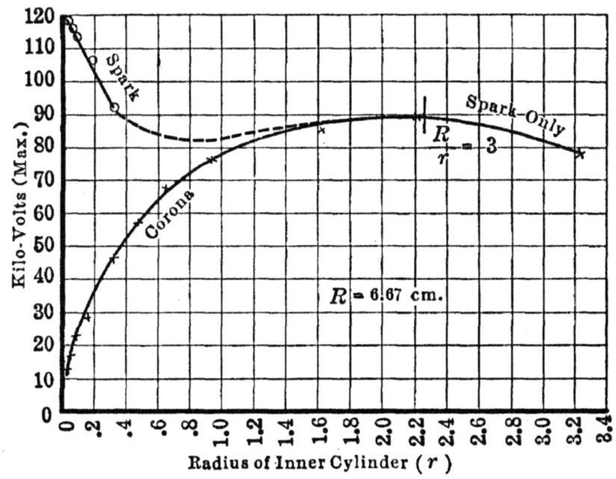

FIG. 91.—Spark-over and corona voltages for concentric cylinders with varying diameter of inner cylinder.

"series resistance," or it grades or distributes the flux density. (See Fig. 91.) Taking the exact equation for parallel wires

$$g_v = \frac{e_v}{r\sqrt{\dfrac{\dfrac{S}{2r} - 1}{\dfrac{S}{2r} + 1}}\cosh^{-1}\dfrac{S}{2r}} \qquad (12b)$$

$$e_v = g_o\left(1 + \frac{0.301}{\sqrt{r}}\right) r \sqrt{\frac{\dfrac{S}{2r} - 1}{\dfrac{S}{2r} + 1}}\cosh^{-1}\frac{S}{2r}$$

Varying r for constant $S = 10$ it is found that e_v is maximum when $S/r = 6.67$. Experiments show this ratio to be 30. This is probably because, at the small spacing, the corona acts as a flexible conductor which collapses and forms a point.

The visual corona voltages, or the spark-over voltages *below* the critical ratio of S/r or R/r, should be of practical value for voltage measurement on account of the accuracy at which they may be determined or calculated for different temperatures, barometric pressures, etc.

Influence on Spark-over of Water and Oil on the Conductor Surface.—Tests with oil and water on the conductor surface were made in a manner exactly similar to the dry spark-over and corona tests. In the oil tests, the surface of the wire was coated with a thin even film by means of an oiled cloth. For the wet tests, water was sprayed on the conductor surface before each reading by means of an atomizer. Figs. 50, 51 and 92 are dry, wet, and oil curves for three different sizes of wire.

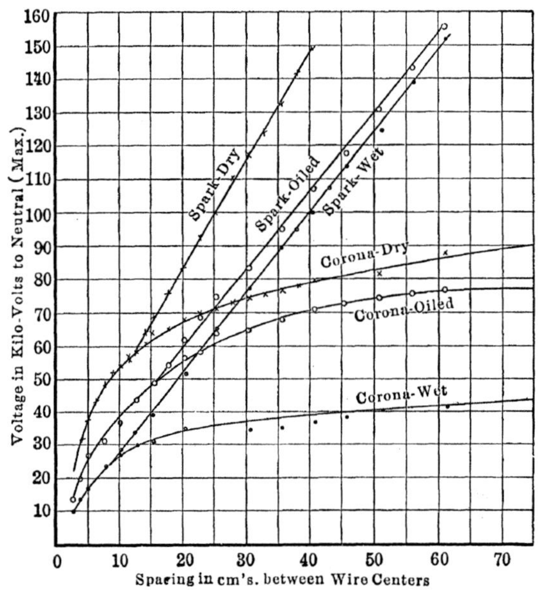

FIG. 92.—Spark-over and visual corona for parallel wires. (Diameter, 0.825 cm. Polished copper. Surfaces dry, wet, and oiled. Maximum volts to neutral given.)

For spark-over both water and oil have approximately the same effect. This curve tends to approach the needle-gap curve.

For corona, water very greatly lowers g_v. Oil lowers g_v but to a much less extent than water. Where the conductor is very small the per cent. increase in diameter due to oil more than compensates for the lowering effect.

The spark gaps which have been useful in measuring high voltages will now be considered.

The Gap as a Means of Measuring High Voltages.—A gap method of measuring high voltages is often desirable in certain commercial and experimental tests. A gap measures the maximum point of the voltage wave and is therefore used in many insulation tests where breakdown also depends upon the maximum voltage. In most commercial tests an accuracy of 2 or 3 per cent. is sufficient. A greater accuracy can be obtained with the sphere gap for special work where special precautions are taken.

The Needle Gap.—The needle gap is unreliable at high voltages because, due to the brush and broken-down air that precedes the spark-over, variations are caused by humidity, oscillations, and frequency.[1]

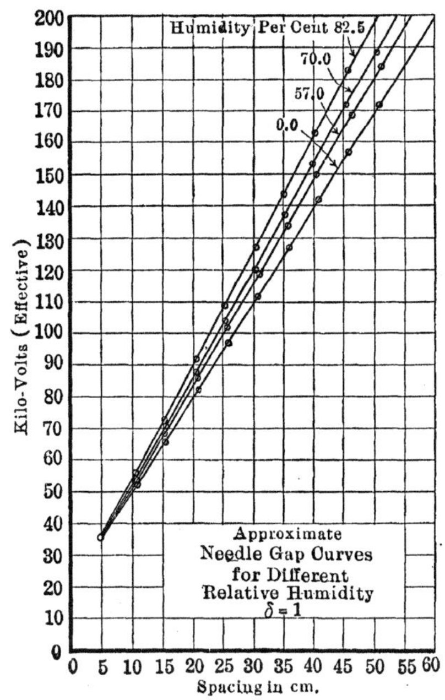

Fig. 93.

The needle gap is also inconvenient because needles must be replaced after each discharge; the spacing becomes very large at high voltages, and the calibration varies somewhat with the sharpness of the needle.

Average Needle Spark-over Voltages

No. 00 Needle, $\delta = 1$

A.I.E.E.

Kv. eff.	Spacing (cm.)	Kv. eff.	Spacing (cm.)
10	1.19	40	6.10
15	1.84	45	7.50
20	2.54	50	9.00
25	3.30	60	11.80
30	4.10	70	14.90
35	5.10	80	18.00

[1] F. W. Peek, Jr., Discussion, A.I.E.E., Feb., 1913.
F. W. Peek, Jr., G. E. Review, May, 1913.

The effect of humidity is shown in Fig. 93, where it can be seen that a higher voltage is required to spark over a given needle gap when the humidity is high than when it is low. (Curves, Fig. 93, are intended only to illustrate this effect.) It is probable that the corona streamers in humid air cause a "fog," and then agglomerate the water particles, which, in effect, increase the size of the electrodes.

All spark-gap curves of whatever form of gap must be corrected for air density—that is, altitude and temperature. For low voltages the spark-over of the needle gap decreases approximately as the air density. At higher voltages the effect becomes more erratic, probably due to humidity.

The Sphere Gap.[1]—The voltage required to spark over a given gap between spheres increases with the diameter of the spheres. Corona cannot form on spheres, or rather, the spark-over point and corona point are coincident if the spacing is not greater than the diameter of the spheres. In practice a spacing as great as three times the radius may be used without appreciable corona. The voltage limit of a given sphere in high-voltage measurements is thus reached when a gap setting greater than three times the radius is required. For accurate work it is preferable to use spacings less than the diameter of the sphere. A larger sphere should then be used. With this space limit the first evidence of stress is complete spark-over; corona can never form, and all of the undesirable effects and variables due to brush discharge and broken-down air are eliminated. Humidity has no measurable effect.

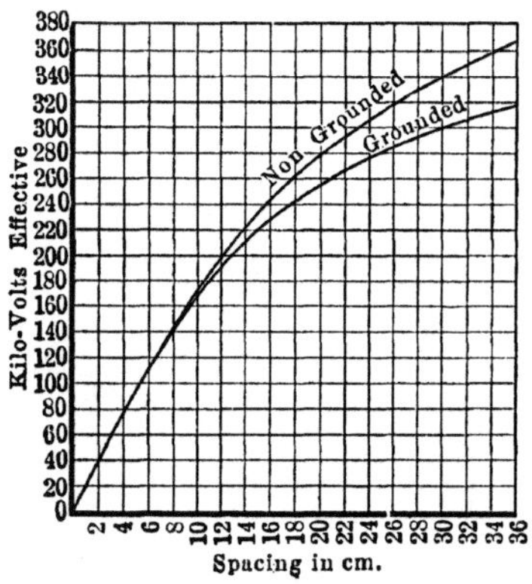

FIG. 94.—Spark-over voltages of 25 cm. spheres.

The space factor is relatively small. Several thousand measurements may be made without repolishing. The curve may be calculated. The only correction is the air-density correction. This has been investigated and the results are given below. The

[1] Chubb and Fortiscue, A.I.E.E., Feb., 1913. "The Calibration of the Sphere Gap Voltmeter."

F. W. Peek, Jr., A.I.E.E., Feb., 1913. "The Sphere Gap as a Means of Measuring High Voltage."

F. W. Peek, Jr., G. E. Review, May, 1913.

TABLE XXI.—SPHERE GAP SPARK-OVER VOLTAGES
6.25-cm. Spheres

Spacing		Kilovolts effective	
Cm.	In.	Non-grounded	Grounded
0.5	0.197	12.0	12.0
1.0	0.394	22.5	22.5
1.5	0.591	31.5	31.5
2.0	0.787	41.0	41.0
3.0	1.181	57.5	56.0
4.0	1.575	70.5	66.0
5.0	1.969	81.0	73.0
6.0	2.362	89.0	79.0
7.0	2.756	96.0	83.0
8.0	3.150	102.0	88.0
9.0	3.543	107.0	90.5
10.0	3.937	110.0	93.0

Each point is the average of five readings. The average variation between maximum and minimum for a given setting is less than 0.5 per cent.

TABLE XXII.—SPHERE GAP SPARK-OVER VOLTAGES
12.5-cm. Spheres

Spacing		Kilovolts effective	
Cm.	In.	Non-grounded	Grounded
0.25	0.098	6.5	6.5
0.50	0.197	12.0	12.0
1.0	0.394	22.0	22.0
1.5	0.591	31.5	31.5
2.0	0.787	41.0	41.0
3.0	1.181	59.0	59.0
4.0	1.575	76.0	75.0
5.0	1.969	91.0	89.0
6.0	2.362	105.0	102.0
7.0	2.756	118.0	112.0
8.0	3.150	130.0	120.0
9.0	3.543	141.0	128.0
10.0	3.937	151.0	135.0
12.0	4.72	167.0	147.0
15.0	5.91	188.0	160.0
17.5	6.88	201.0	168.0
20.0	7.87	213.0	174.0

correction is quite simple. Fig. 94 gives typical sphere-gap curves for both spheres insulated and for one sphere grounded. Tables XXI to XXV give spark-over curves for 6.25, 12.5, 25, 50 and 100 cm. spheres at sea level (25 deg. C.—76 cm. barometer). $\delta = 1$.

TABLE XXIII.—SPHERE GAP SPARK-OVER VOLTAGES

25-cm. Spheres

Spacing		Kilovolts effective	
Cm.	In.	Non-grounded	Grounded
0.5	0.197	11	11
1.0	0.394	22	22
1.5	0.591	32	32
2.0	0.787	42	42
2.5	0.983	52	52
3.0	1.181	61	61
4.0	1.575	78	78
5.0	1.919	96	94
6.0	2.362	112	110
7.5	2.953	135	132
10.0	3.937	171	166
12.5	4.92	203	196
15.0	5.91	230	220
17.5	6.88	255	238
20.0	7.87	278	254
22.5	8.85	297	268
25.0	9.83	314	280
30.0	11.81	339	300
40.0	15.75	385	325

Calculation of Curves.—The gradient or stress on the air at the sphere surface, where it is greatest, is found mathematically

$$g = \frac{e}{X} f \text{ kv./cm.} \tag{13}$$

Where e is the applied voltage in kilovolts, X is the spacing in centimeters, f is a function of X/R and R is the radius of the sphere in centimeters.

TABLE XXIV.—SPHERE GAP SPARK-OVER VOLTAGES

50-cm. Spheres

Spacing		Kilovolts effective	
Cm.	In.	Non-grounded	Grounded
2	0.787	40.0[1]	40
4	1.575	76.5	76
6	2.362	115.5	112
8	3.150	149.0	145
10	3.937	189.0	185
12	4.72	224.3	220
14	5.51	255.5	250
16	6.30	285.0	275[2]
20	7.87	335.0	320
25	9.83	393.0	377
30	11.81	445.0	420
35	13.80	493.0	456
40	15.75	537.0	489
45	17.72	573.0	516
50	19.19	605.0	541
55	21.65	633.0	561
60	23.62	660.0	579
65	25.60	684.0	594
70	27.56	705.0	608
75	29.55	725.0	619

[1] These values are calculated.
[2] Spacings above 16 cm. are calculated.

Then

$$g_s = \frac{e_s}{X} f \text{ kv./cm.}$$

where e_s is the spark-over voltage and g_s is the apparent strength of air.

f is found mathematically and tabulated on page 27, for the non-grounded and grounded cases. For the non-grounded case we have found experimentally that g_s, the apparent surface gradient at spark-over, increases with decreasing radius of sphere, as g_v for corona on wires increases for decreasing radius of wire.

TABLE XXV.—SPHERE GAP SPARK-OVER VOLTAGES

100-cm. Spheres

These values are calculated

Spacing		Kilovolts effective	
Cm.	In.	Non-grounded	Grounded
1.0	0.394	20	20
3.0	1.181	60	60
5.0	1.969	100	100
10.0	3.937	195	195
15.0	5.91	283	280
20.0	7.87	364	360
30.0	11.81	520	505
40.0	15.75	650	615
50.0	19.69	770	730
60.0	23.62	870	810
70.0	27.56	956	895
80.0	31.50	1044	956
90.0	35.43	1107	1010
100.0	39.37	1182	1057
110.0	43.35	1238	1090
120.0	47.20	1290	1133
130.0	51.20	1335	1160
140.0	55.70	1378	1189
150.0	59.10	1412	1212

For a given size of sphere, g_s is practically constant, independent of spacing, between the limits of $X = 0.54\sqrt{R}$ and $X = 2R$.

The *average* gradient between these limits of separation is

$$g_s = 27.2\left(1 + \frac{0.54}{\sqrt{R}}\right)\text{kv./cm. max.} \qquad (a)$$

$$g_s = 19.3\left(1 + \frac{0.54}{\sqrt{R}}\right)\text{kv./cm. eff. sine wave.}^1 \qquad (b)$$

The maximum variation from the average between the limits may be 2 per cent. When X is less than $0.54\sqrt{R}$, g_s increases very rapidly because the spacing is then less than the "rupturing energy

[1] F. W. Peek, Jr., "Law of Corona III," A.I.E.E., June, 1913.

distance."[1] Above $X = 3R$, g_s apparently gradually increases. This increase seems only apparent and due to the shanks, surrounding objects, etc., better distributing the flux or lessening the flux density. When both spheres are insulated and of practical size, the change is not great within the prescribed limits. In this case the neutral of the transformer should be grounded so that spheres are at equal and opposite potential. When one sphere is grounded, however, this apparent increase of gradient is very great if the mathematical f, which does not take account of the effect of surrounding objects, is used. For this reason f was found experimentally, assuming g_s constant within the limits, as it is in the non-grounded case, and finding values of f_o corresponding to the different values of X/R. Any given value of the ratio X/R should require a constant f_o to keep g_s constant independent of R. This was found to check.[2]

The curves may be approximately calculated thus:

$$e_s = g_s \frac{x}{f} \begin{array}{l}\text{(non-grounded)} \\ \text{effective sine wave.}\end{array} \qquad (13a)$$

$$e_s = g_s \frac{x}{f_o} \begin{array}{l}\text{(grounded)} \\ \text{effective sine wave.}\end{array} \qquad (13b)$$

Where g_s is calculated from the equation (b), and f or f_o are found from the table on page 27 for the given X/R. These equations have been given for theoretical rather than practical reasons. *Curves should be calculated only when standard measured curves cannot be obtained.* Measured curves are given here. The average error, however, for curves calculated from the above equations, for 2-cm. diameter spheres and over, should not be greater then 2 per cent. The accuracy of calculations is not as great as in the case of the starting point of corona on wires.

The Effect of Air Density or Altitude and Temperature: Correction Factor. Practical Application.—We have found that the average gradient for various air densities may be expressed

$$g_s = 27.2\delta\left(1 + \frac{0.54}{\sqrt{\delta R}}\right)\text{kv./cm. max.}$$

$$g_s = 19.3\delta\left(1 + \frac{0.54}{\sqrt{\delta R}}\right)\text{kv./cm. effective.}$$

where δ is the relative air density. (See page 51.)

[1] F. W. Peek, Jr., "Law of Corona III," A.I.E.E., June, 1913.

[2] f_o was determined with the grounded sphere 4 to 5 diameters above ground. In practice, this may vary from 4 to 10 diameters without great error. See Table XXXIV. Voltage values in tables correspond to 4 to 5 diameters above the ground for this case.

The standard curve may be made to apply to any given altitude by multiplying the standard curve voltage at different spacings by the correction factor thus

$$e_1 = e\left\{\frac{19.3\delta\left(1 + \frac{0.54}{\sqrt{\delta R}}\right)}{19.3\left(1 + \frac{0.54}{\sqrt{R}}\right)}\right\}$$

$$= e\sqrt{\delta}\left\{\frac{\sqrt{\delta R} + 0.54}{\sqrt{R} + 0.54}\right\} = ea$$

TABLE XXVI

Approximate corresponding altitude	Barometer		Values of a at 25 deg. C. for standard spheres of the following diameter, cm.						
	Cm., Hg	In., Hg							
Ft.			6.25	12.5	25.0	37.5	50.0	75.0	100.0
0	76.00	29.92	1.000	1.000	1.000	1.000	1.000	1.000	1.000
500	74.58	29.36	0.981	0.980	0.980	0.979	0.979	0.979	0.979
1,000	73.14	28.79	0.964	0.963	0.962	0.961	0.960	0.960	0.960
1,500	71.77	28.25	0.948	0.946	0.945	0.944	0.943	0.942	0.942
2,000	70.42	27.72	0.932	0.929	0.927	0.926	0.925	0.924	0.924
2,500	60.09	27.20	0.916	0.913	0.911	0.909	0.908	0.907	0.907
3,000	67.74	26.67	0.902	0.899	0.897	0.895	0.893	0.892	0.891
3,500	66.51	26.18	0.887	0.884	0.882	0.880	0.878	0.876	0.875
4,000	65.25	25.69	0.873	0.870	0.867	0.865	0.863	0.861	0.860
4,500	64.01	25.23	0.859	0.855	0.852	0.850	0.848	0.846	0.845
5,000	62.79	24.72	0.845	0.841	0.838	0.835	0.833	0.831	0.830
6,000	60.45	23.80	0.817	0.812	0.808	0.805	0.803	0.801	0.800
7,000	58.22	22.93	0.791	0.786	0.782	0.779	0.776	0.774	0.772
8,000	56.03	22.05	0.765	0.759	0.754	0.750	0.748	0.746	0.744
9,000	53.84	21.20	0.739	0.733	0.728	0.724	0.721	0.719	0.717
10,000	51.85	20.41	0.716	0.709	0.703	0.698	0.694	0.692	0.690
12,000	48.09	18.93	0.669	0.661	0.656	0.651	0.647	0.644	0.642
15,000	42.88	16.88	0.606	0.596	0.589	0.585	0.580	0.577	0.575

In order to avoid the trouble of calculating in practice, the factor is tabulated in Tables XXVI and XXVII. This correction is very accurate. Table XXVI gives the correction factor for different sizes of spheres at different barometric pressure and at constant temperature. When the voltage strikes across a given gap the voltage, e, corresponding to the gap is found from the standard curve and multiplied by the correction factor a, or a curve may be plotted corresponding to a given barometric pressure.

Thus
$$e_1 = ea$$

Table XXVII gives the correction factor for various values of δ. δ may be calculated for the given temperature and barometric pressure and correction factor then found from the table. Fig. 95 gives the standard curve for the 25-cm. sphere (non-grounded) (25 deg. C., 76 cm. bar. pressure) and curves calculated therefrom for 25 deg. C. and various barometric pressures.

TABLE XXVII.—CALCULATED VALUES OF a FOR DIFFERENT VALUES OF δ

$$a = \sqrt{\delta}\left\{\frac{\sqrt{\delta R} + 0.54}{\sqrt{R} + 0.54}\right\}$$

Relative air density	Values of a Diameter of standard spheres in cm.						
δ	6.25	12.5	25.0	37.5	50.0	75.0	100.0
0.50	0.547	0.535	0.527	0.522	0.519	0.517	0.516
0.55	0.594	0.583	0.575	0.570	0.567	0.565	0.564
0.60	0.640	0.630	0.623	0.618	0.615	0.613	0.612
0.65	0.686	0.677	0.670	0.665	0.663	0.661	0.660
0.70	0.732	0.724	0.718	0.714	0.711	0.709	0.708
0.75	0.777	0.771	0.766	0.762	0.759	0.757	0.756
0.80	0.821	0.816	0.812	0.809	0.807	0.805	0.804
0.85	0.866	0.862	0.859	0.857	0.855	0.854	0.853
0.90	0.910	0.908	0.906	0.905	0.904	0.903	0.902
0.95	0.956	0.955	0.954	0.953	0.952	0.951	0.951
1.00	1.000	1.000	1.000	1.000	1.000	1.000	1.000
1.05	1.044	1.045	1.046	1.047	1.048	1.049	1.049
1.10	1.092	1.092	1.094	1.095	1.096	1.097	1.098

Experimental Determination of the Effect of Air Density.—The equation for the air density correction factor was determined by an extensive investigation of the spark-over of spheres in a large wooden cask arranged for exhaustion of air. This cask was built of paraffined wood and was 2.1 meters high by 1.8 meters in diameter inside. (See Fig. 96.)

Tests were made by setting a given size of sphere at a given spacing, gradually exhausting the cask, and reading spark-over

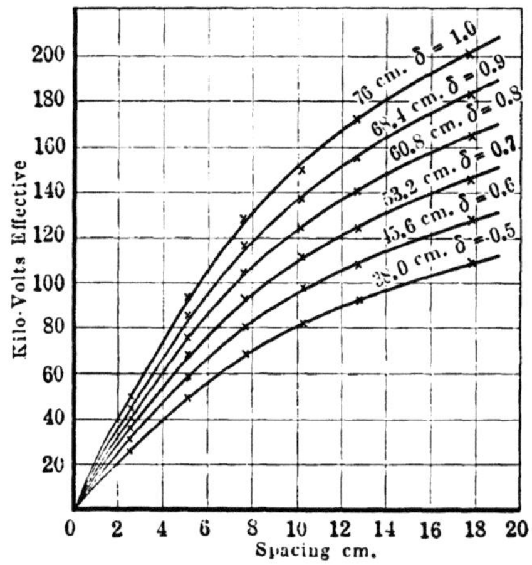

FIG. 95.—Sphere gap spark-over voltages at various air pressures. (12.5 cm. spheres. Non-grounded. Curves calculated. Points measured.)

voltage at intervals as the air pressure was changed. (Temperature was always read, but varied only between 16 deg. and

FIG. 96.—Cask for the study of the variation of spark-over and corona voltages with air pressure.

21 deg. C.) This was repeated for various spacings on spheres ranging in diameter from 2 cm. to 25 cm. At the start, the

possible effect of spark-overs on the succeeding ones in the cask was investigated and found to be nil or negligible. A resistance of 1 to 4 ohms per volt was used in series with the spheres. Wave shape was measured and corrected for. Voltage was read on a voltmeter coil, by step-down transformer and by ratio. Precautions were taken as noted in other chapters.

In order to illustrate the method of recording data, etc., a small part of the data for various spheres and spacings is given in Tables XXVIII to XXXII. Considerable data are plotted in curves, Figs. 97 to 99. The points are measured values. The drawn lines are calculated by multiplying the voltage values from the standard curves at $\delta = 1$, by the correction factor.

TABLE XXVIII.—SPHERE GAP SPARK-OVER VOLTAGES AND GRADIENT

2.54-cm. Spheres. Non-grounded

Barometer 75.5

Spacing		Temp.	Press., cm. of Hg	Relative air density	Kv.		g_s measured	
Cm.	In.				Eff.	Max.	Eff.	Max.
0.635	0.25	15°	75.5	1.028	15.7	22.2	29.2	41.3
......	......	......	72.6	0.990	15.3	21.6	28.5	40.3
......	......	......	70.5	0.957	14.9	21.1	27.7	39.2
......	......	......	68.4	0.930	14.4	20.4	26.8	38.0
......	......	......	65.8	0.895	14.0	19.8	26.0	36.8
......	......	......	62.6	0.852	13.6	19.2	25.3	35.7
......	......	......	60.0	0.818	13.2	18.7	24.6	34.8
......	......	......	56.7	0.776	12.5	17.2	23.2	32.0
......	......	......	54.8	0.745	12.1	17.1	22.5	31.8
......	......	......	52.6	0.717	11.8	16.7	22.0	31.1
......	......	......	50.5	0.686	11.3	16.0	21.0	29.8
......	......	......	47.6	0.649	10.7	15.2	20.0	28.3
......	......	......	45.4	0.618	10.4	14.7	19.3	27.4
......	......	......	42.4	0.576	9.8	13.8	18.2	25.8
......	......	......	40.0	0.544	9.3	13.1	17.2	24.4
......	......	......	36.8	0.500	8.7	12.2	16.1	22.8

TABLE XXIX.—SPHERE GAP SPARK-OVER VOLTAGES AND GRADIENTS
5.08 Spheres. Non-grounded

Spacing		Relative air density	Kv.		g_s measured	
Cm.	In.		Effective	Maximum	Effective	Maximum
5.08	2	1.018	78.4	111.0	27.4	38.9
5.08	2	0.980	75.6	107.0	26.5	37.6
5.08	2	0.944	73.1	103.5	25.6	36.2
5.08	2	0.903	70.7	100.0	24.8	35.0
5.08	2	0.872	68.7	97.2	24.0	34.0
5.08	2	0.836	64.0	90.5	22.4	31.7
5.08	2	0.798	63.6	90.0	22.3	31.5
5.08	2	0.764	61.1	86.4	21.4	30.2
5.08	2	0.726	58.9	83.3	20.6	29.2
5.08	2	0.682	56.1	80.5	19.6	28.2
5.08	2	0.654	54.0	76.4	18.9	26.8
5.08	2	0.618	51.5	73.0	18.0	25.6
5.08	2	0.578	49.0	69.3	17.2	24.3
5.08	2	0.544	45.8	64.7	16.0	22.6
5.08	2	0.510	43.3	61.2	15.2	21.4

TABLE XXX.—SPHERE GAP SPARK-OVER VOLTAGES AND GRADIENTS
12.5-cm. Spheres. Non-grounded

Spacing		Relative air density	Kv.		g_s measured	
Cm.	In.		Effective	Maximum	Effective	Maximum
12.7	5	0.982	163.0	230.0	23.1	32.7
12.7	5	0.951	156.0	221.0	22.2	31.4
12.7	5	0.917	150.0	212.5	21.3	30.2
12.7	5	0.880	147.0	208.0	20.9	29.5
12.7	5	0.846	143.5	203.0	20.4	29.8
12.7	5	0.807	139.5	197.5	19.8	28.0
12.7	5	0.780	134.5	190.0	19.1	27.0
12.7	5	0.736	131.0	185.5	18.6	26.3
12.7	5	0.699	125.0	177.0	17.7	25.1
12.7	5	0.666	120.5	170.0	17.1	24.2
12.7	5	0.637	115.5	163.0	16.4	23.1
12.7	5	0.598	109.5	155.0	15.5	22.0
12.7	5	0.561	104.5	147.5	14.8	21.0
12.7	5	0.541	101.0	142.5	14.3	20.2

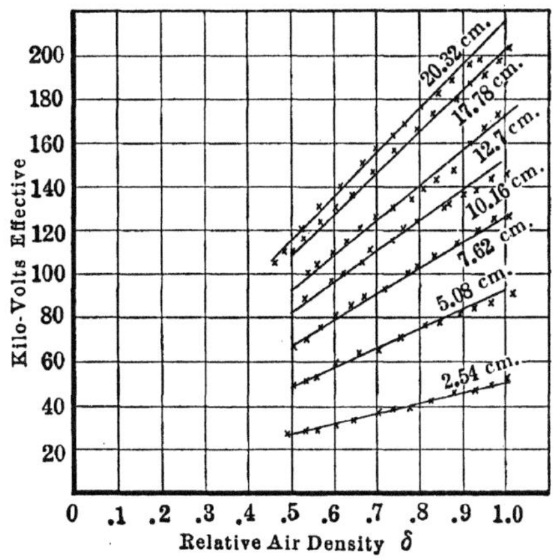

FIG. 97.—Spark-over voltages at different air densities. (12.5 cm. spheres non-grounded. Figures on curves denote spacing.)

TABLE XXXI.—SPHERE GAP SPARK-OVER VOLTAGES AND GRADIENTS

12.5-cm. Spheres. Grounded

Spacing		Relative air density	Kv.		g_s measured	
Cm.	In.		Effective	Maximum	Effective	Maximum
6.35	2.5	0.908	95.5	135.0	21.2	30.0
6.35	2.5	0.869	92.5	130.5	20.5	29.0
6.35	2.5	0.828	88.6	125.0	19.7	27.8
6.35	2.5	0.796	85.8	121.0	19.0	26.9
6.35	2.5	0.758	81.1	114.5	18.0	25.4
6.35	2.5	0.723	78.2	110.5	17.3	24.5
6.35	2.5	0.690	73.2	103.5	16.2	23.0
6.35	2.5	0.653	71.5	101.0	15.9	22.4
6.35	2.5	0.620	68.2	96.3	15.1	21.4
6.35	2.5	0.582	64.3	90.9	14.2	20.2
6.35	2.5	0.539	60.7	85.8	13.5	19.0
6.35	2.5	0.439	55.6	78.5	12.3	17.4

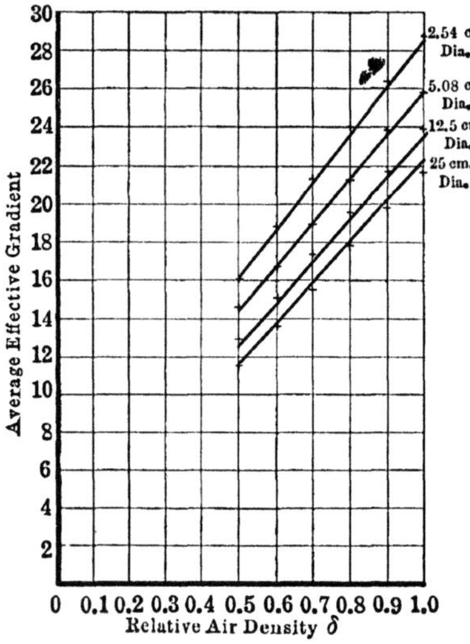

FIG. 98.

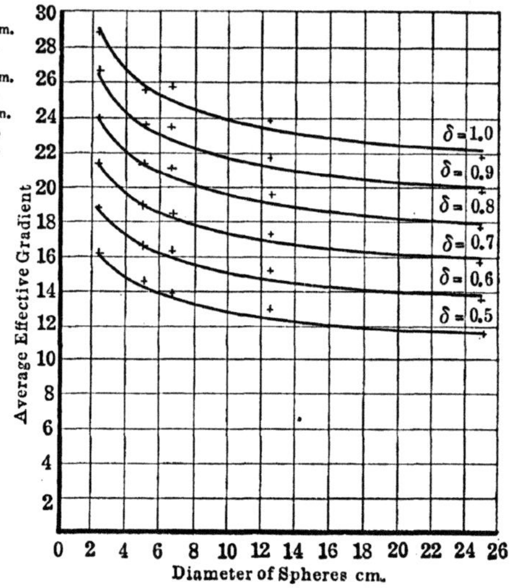

FIG. 99.

Spark-over gradients at different air densities for several sizes of spheres. (Points measured. Curves calculated from $g_s = 19.3\delta\left(1 + \frac{0.54}{\sqrt{\delta R}}\right)$)

TABLE XXXII.—SPHERE GAP SPARK-OVER VOLTAGES AND GRADIENTS

25-cm. Spheres. Non-grounded

Spacing		Relative air density	Kv.		g_s measured	
Cm.	In.		Effective	Maximum	Effective	Maximum
7.62	3	1.018	139.0	196.5	22.2	31.5
7.62	3	0.978	133.5	189.0	21.4	30.3
7.62	3	0.942	129.5	183.0	20.8	29.3
7.62	3	0.906	126.0	178.0	20.2	28.5
7.62	3	0.888	121.5	172.0	19.5	27.6
7.62	3	0.839	115.0	163.0	18.4	26.1
7.62	3	0.796	111.0	157.0	17.8	25.2
7.62	3	0.752	105.5	149.0	16.9	23.9
7.62	3	0.718	101.5	142.0	16.3	22.7
7.62	3	0.685	96.2	136.0	15.4	21.8
7.62	3	0.646	91.5	129.5	14.6	20.7
7.62	3	0.608	87.0	123.0	13.9	19.7
7.62	3	0.570	81.3	115.0	13.0	18.4
7.62	3	0.527	74.7	105.5	12.0	16.9
7.62	3	0.491	70.8	100.0	11.3	16.0

The calculated values check the measured values closely.

The equation for the correction factor was deduced from measured values as follows:

From a former investigation it was found that at $\delta = 1$ the average gradient

$$g_s = g_o\left(1 + \frac{0.54}{\sqrt{R}}\right)^{1}$$

From this investigation it was found that the average gradient at various values of δ is

$$g_s = g_o\delta\left(1 + \frac{0.54}{\sqrt{\delta R}}\right)$$

TABLE XXXIII.—AVERAGE EFFECTIVE RUPTURING GRADIENT FOR SPHERES OF SEVERAL DIAMETERS AND VARYING AIR DENSITIES

Diameter of Spheres, cm. Surface Gradients

Columns marked "Calc." are from, $g_s = 19.3\delta\left(1 + \frac{0.54}{\sqrt{\delta R}}\right)$

δ	2.54		5.08		12.5		25	
	Meas.	Calc.	Meas.	Calc.	Meas.	Calc.	Meas.	Calc.
1.00	28.7	28.5	25.8	25.8	24.0	23.4	21.9	22.2
0.90	26.2	26.1	23.7	23.3	21.8	21.2	19.9	20.1
0.80	24.0	23.7	21.3	21.3	19.6	19.1	17.9	18.0
0.70	21.3	21.2	19.0	19.0	17.4	16.9	15.7	15.9
0.60	18.7	18.7	16.7	16.6	15.2	14.7	13.6	13.8
0.50	16.1	16.1	14.6	14.3	13.0	12.5	11.6	11.7

The average measured gradients for various values of δ are given in Table XXXIII, the calculated values from the equation are also given. The check is quite close. It should be remembered, however, that these are average values over this range of spacing and that there is a small variation at different spacings as already explained. (See Figs. 98 and 99.)

Precautions against Oscillations in Testing.—A non-inductive resistance of 1 or 4 ohms per volt should always be placed directly in series with the gap. For the non-grounded gap, one-half should be placed on each side. When one gap is grounded all of the resistance should be placed on the insulated side. One

[1] F. W. Peek, Jr., "Law of Corona III," A.I.E.E., June, 1913.
F. W. Peek, Jr., Discussion, A.I.E.E., Feb., 1913.

7

object of the resistance is to prevent oscillations from the test piece, as a partial arc-over on a line insulator, reaching the gap. Another object is to limit the current discharge. This resistance is of special importance when tests are being made on apparatus containing inductance and capacity. If there is no resistance, when the gap sparks over, oscillations will be produced which will cause a very high local voltage rise over parts of the winding. If sufficient resistance is used these oscillations will be damped out. This is illustrated in Fig. 100, which shows results of a test on a high-voltage transformer.

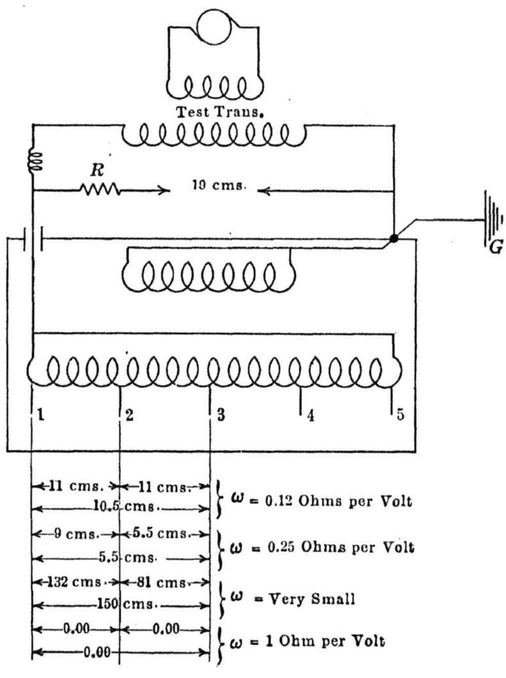

FIG. 100.

Referring to Fig. 100, the high-voltage winding of the transformer under test is short circuited and connected to one terminal of the testing transformer. The other side of the testing transformer is grounded. The low-voltage winding of the transformer under test is short circuited, connected to the case and ground. Voltage is gradually applied to the transformer under test until the "measuring gap" sparks over. Insulated taps, 1, 2, 3, 4, 5, are brought out at equally spaced points on the high-tension winding of the transformer under test. Auxiliary needle gaps are placed between 1 and 2, 2 and 3, and 1 and 3, to measure the voltage which appears across these sections of the winding when the main measuring gap discharges. The numbers between 1–2, 2–3, and 1–3 represent the sparking distances of the local voltages caused by a discharge of the measuring gap. Four cases are given with different values of resistance ω in the main gap. When $\omega = 1$ ohm per volt, the local oscillations are completely damped out.

With small resistance in the gap, a 19-cm. spark-over causes a voltage to build up between coils 1 and 3 (which sparks over a 150-cm. gap), although the total applied voltage across the transformer is only equivalent to a 19-cm. gap. The apparatus may thus be subjected to strains far beyond reason, and either broken down or very much weakened. Water-tube resistance is the most reliable. A metallic resistance, if non-inductive and of small capacity, may be used. Carbon or graphite rods should be avoided as, although they may measure up to a very high resistance at low voltage, the resistance may become very low at high voltage by "coherer" action.

When the tested apparatus is such that there is considerable incipient arcing before spark-over, it is better to use the sphere to determine the "equivalent" ratio of the transformer at a point in voltage below the voltage at which this arcing occurs. The sphere gap should then be widened out, the spark-over voltage measured on the low-voltage side of the transformer or in the voltmeter coil, and multiplied by this equivalent ratio. It must also be remembered that resistances do not dampen out low frequency surges resulting from a short circuit, etc.

Miscellaneous Precautions.—In making tests it is desirable to observe the following precautions:

The shanks should not be greater in diameter than one-fifth the sphere diameter. Metal collars, etc., through which the shanks extend should be as small as practicable, and not come closer to the sphere than the gap distance at maximum opening. The effect of a large plate or plates on the shanks is given in Table XXXIV. The sphere diameter should not vary more than 0.1 per cent., and the curvature, measured by a spherometer, should not vary more than 1 per cent. from that of a true sphere of the required diameter. The spheres should be at least twice the gap setting from surroundings. This is especially important if the objects are large conducting, or semi-conducting masses, walls, floor, etc. Care must also be taken to so place the spheres that external fields are not superposed upon the sphere gap. This is likely to result, especially in the nongrounded case, from a large mass of resistance units or connecting leads, etc., in back of and in electrical connection with either sphere. The error may be plus or minus as indicated for the small plates in Table XXXIV. With the water tube resistance this condition is not likely to obtain as the tube may be brought directly to the sphere as an extension of the shank. Great precautions

TABLE XXXIV.—EFFECT OF METAL PLATES ON THE SHANKS. DISTANCE TO GROUND ON ARC-OVER—OF 6.25-CM. SPHERES

Per Cent. Change of Voltage

Sphere gap, cm.	Non-grounded 5 cm. diameter plates. 6.25 cm. Back of		Grounded 5 diameter plates. 6.25 cm. Back of	
	Both spheres	One sphere	Insulated sphere	Grounded sphere
1.5	0.0	0.0	+ 0.7	−0.7
3.0	+1.0	−1.0	+ 1.5	−1.5
6.0	+2.0	−2.0	+ 3.0	−2.0

APPROXIMATE EFFCT OF DISTANCE ABOVE GROUND WHEN ONE SPHERE IS GROUNDED

Diameters of grounded sphere above ground	Percentage variation from standard curves for different spacings		
	$x = 2R$	$x = R$	$x = \frac{R}{2}$
0	− 10.0	− 5.5	0.0
1	− 4.5	− 3.0	0.0
2	− 2.0	− 1.0	0.0
3	− 1.0	− 0.5	0.0
4	− 0.0	− 0.0	0.0
5	+ 0.5	+ 0.3	0.0
6	+ 1.0	+ 0.5	0.0
10	+ 2.5	+ 1.0	0.0
20	+ 2.5	+10.0	0.0

NOTE: The spacing, X, is given in terms of radius, R, to make the correction applicable to any size of sphere. The distance of the grounded sphere above ground is, for the same reason, given in terms of the sphere diameter. The (+) sign means that a higher voltage is required to arc over the gap than that given by the standard curve; the (−) sign indicates that a lower voltage is required. The standard curves are made with the grounded spheres from 4 to 5 diameters above ground. In practice it is desirable to work between 4 and 10 diameters, never under 3. Above 10 the variations in per cent. error remain about the same.

When both spheres are insulated, with the transformer neutral at the mid point, there is practically no variation in voltage for different distances above ground.

are necessary at very high voltages to prevent leakage over stands, supports, etc., and to prevent corona and brush discharges. Unless such precautions are taken errors will result.

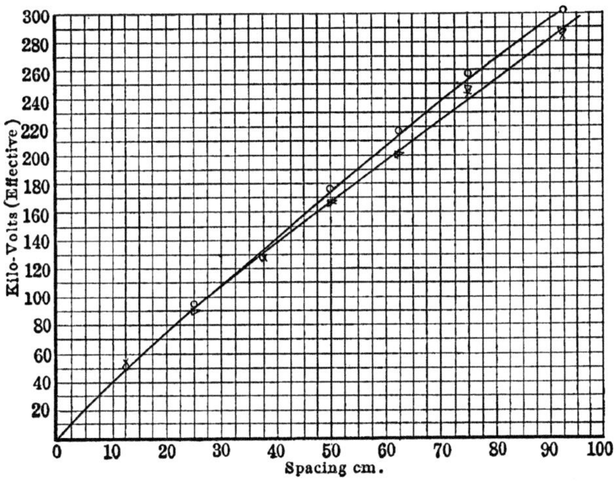

FIG. 101.—Spark-over voltages between points. o, dry; Δ, wet; x, rain. 0.25″ per minute. 60° points on 1.25 cm. rods.

Rain and Water on Sphere Surface.—Figs. 101 and 102 show the effect of rain and water on points and spheres. The ratio

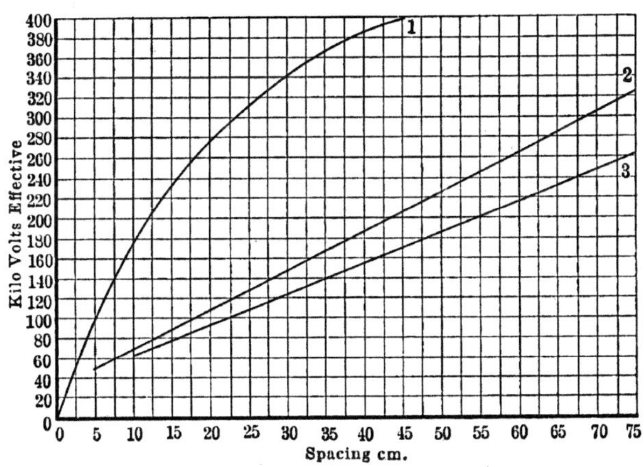

FIG. 102.—Spark-over voltages between 25 cm. spheres. 1, dry; 2, wet surface; 3, rain (0.25″ per min.).

of dry to rain (0.2 in. per min.) spark-over voltage for a given spacing will average about 2.5 for 6.25 to 50-cm. spheres.

High Frequency, Oscillations, Impulses.—At the present time a great deal is said of the effects of "high frequency" on insulation without differentiating between continuous sine wave high frequency, oscillations, and steep wave front impulses. This has caused considerable confusion, as the effects may be quite different, and are all attributed to the same cause.

High frequency from an alternator, or a series of oscillations, may cause high-insulation loss, heating, and the resulting weakening of the insulation. Single trains of oscillations, or single impulses, may not produce heating. Energy, and therefore definite finite time, is required to rupture insulation. For a single impulse, or oscillation, where the time is limited, a greater voltage should be required to break down a given insulation than for continuous high frequency or for low frequency. Experiments

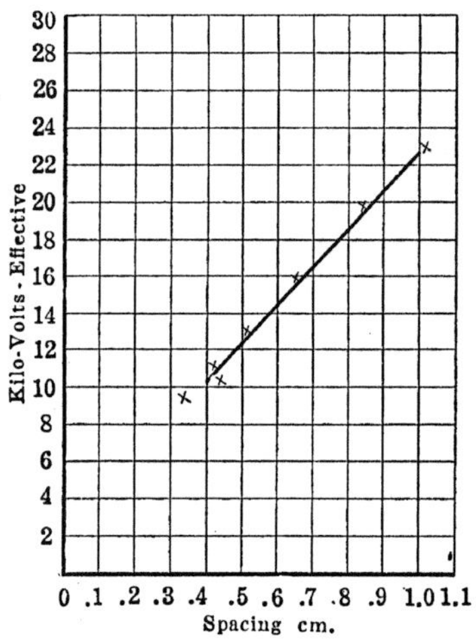

FIG. 103.—Sphere gap spark-over voltages at 60 cycles and 1000 cycles. 5.08 cm. spheres.

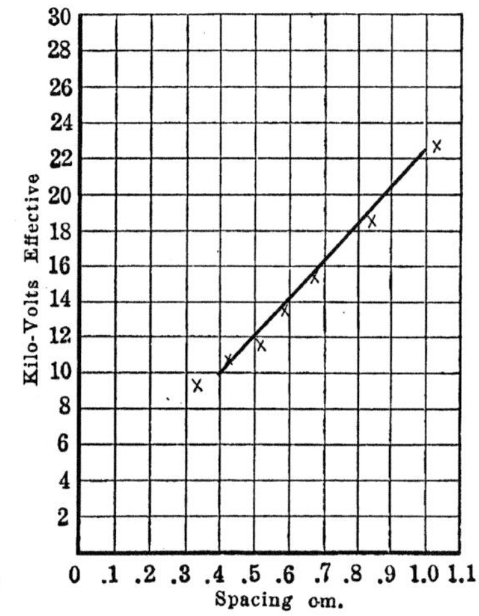

FIG. 104.—Sphere gap spark-over voltages at 60 cycles and 1000 cycles. 12.5 cm. spheres.

(Drawn curves 60 cycles. Points 1000 cycles.)

bear this out. High local voltages may result from high frequency, oscillations, etc. The flux may lag behind the voltage in non-homogeneous insulations. (This does not apply to air.) For a given thickness of a homogeneous insulation, and when heating does not result, a greater oscillatory or impulse voltage of short duration is generally necessary to cause puncture than at 60 cycles.

Some of the effects will now be discussed:

Frequency.—Over the commercial range there is no variation in sphere-gap voltages due to frequency. Figs. 103 and 104 show spark-over curves up to 25 kv. at 1000-cycle sine wave from an alternator. The voltage was measured by a static voltmeter calibrated at 60 cycles. The drawn curve is the 60-cycle curve, the points are measured values. Fig. 105 gives a 60-cycle curve, and also a 40,000-cycle curve from a sine wave alternator. The voltage in this case was measured by a static voltmeter. No special care was taken to polish the sphere surfaces. At low frequencies, at rough places on the electrode surface, there is local overstress; but even if the air is broken down, the loss at these places is very small and the streamers inappreciable. At continuous high frequency, say 40,000 cycles, a local breakdown at a rough point probably takes place at very nearly the same gradient as at 60 cycles, but the energy loss after the breakdown at this point occurs may be 1000 times as great. This forms a needle-like streamer which increases the stress and local loss. Spark-over then takes place from the "electric needle" at a lower voltage than the true sphere-gap voltage; thus it seems that the air at high frequency *of the above order* is only apparently of less strength. These "electric needles" when once formed may be blown to different parts of the sphere surface. The corona starting point appears to take place at a lower voltage at high frequency, because the local loss at rough points, which occurs before the true critical voltage is reached, is very high at high frequency and distorts the field and masks the true starting voltage. The loss at rough points starts at the same voltage at low frequency but is inappreciable and cannot change conditions. If the sphere surfaces are very highly polished it seems that the high-frequency spark-over voltage should check closely with the 60-cycle voltage. This should also apply for corona on polished wires. The following limitation, however, applies to both cases. At *continuous high frequency* when the rate of energy or power is great, frequency may enter into the energy distance equation thus

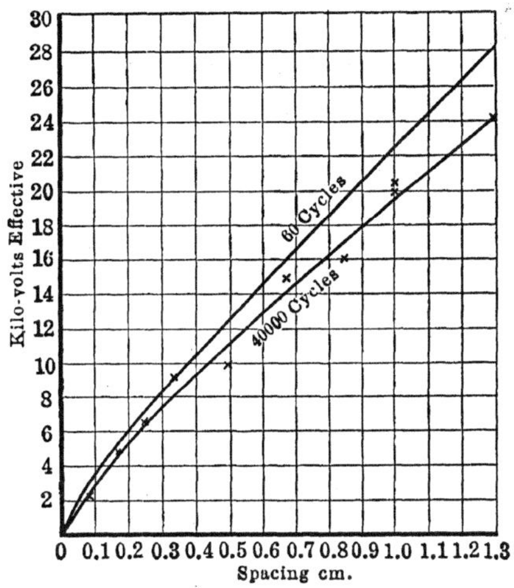

FIG. 105.—Sphere gap spark-over voltages at 60 cycles and 40,000 cycles.

$$A\sqrt{\frac{r}{\delta}}\,\phi\,(f)$$

and spark-overs take place at lower voltages at very high frequency. This is more fully discussed in Chapter VIII.

Destruction of insulation by high frequency, when heating does not result, is due to local overvoltage. For instance, low high-frequency voltage may be applied to a piece of apparatus containing inductance and capacity, as a transformer. On account of the capacity and inductance, very high overvoltage may be built up and breakdown result due to overvoltage. The petticoat of an insulator may be broken down by an "electric needle," forming as described above, bringing the total stress on the thin petticoat. The sphere curves check very closely for oscillatory voltage of short duration and voltages of *steep wave front,* even under ordinary conditions of surface. With needle gaps the results are quite different at high and low frequency. At continuous high frequency the needle point becomes hot, due to the loss, and spark over takes place at low voltage. For impulses, the opposite is true. The sphere gap is thus the most reliable means of measuring oscillatory voltages, and voltages of steep wave front. When used for continuous high frequency the surfaces must be kept highly polished.

Dielectric Spark Lag, Steep Wave Front, and Oscillations.—A given voltage is required to spark over a given gap when the time is not limited. The time necessary to supply the rupturing energy to the gap or surface depends upon the shape and length of the gap or surface, and the initial condition—as somewhat on[1] initial ionization, rate of application of voltage, etc. There is a given minimum time at which the rupturing energy may be supplied at any voltage. If the time is limited, as by steep wave front, a higher voltage is required to accomplish the same results.

If the voltage required to spark over a given surface or gap is measured at 60 cycles and at steep wave front, it is found that the latter voltage is higher. The ratio of impulse spark-over voltage to the 60-cycle spark-over voltage is here termed the "impulse ratio." Spheres measure very closely the actual voltage, even at steep impulses. Test results made by the author showing the increase in voltage for the needle gap are given in Table XXXV. The impulse used was not steep enough to appreciably affect the sphere. The impulse voltage was increased until spark-over took place on the test piece. The voltage was then measured by the sphere (not in multiple with test piece). Note the very high impulse voltage required to

[1] See page 198.

TABLE XXXV.—IMPULSE RATIO OF NEEDLE GAP

Spacing of needles (cm.)	Voltage required to cause needle gap to spark over		Ratio impulse to 60 cycles, voltage
	60 cycles (max.)	Impulse measured by 25-cm. spheres (dia.) (max.)	
4.3	43.1	62.0	1.45
8.0	64.8	96.1	1.49
12.0	85.0	140.0	1.65
25.0	134.0	260.0	1.94

TABLE XXXVI.—IMPULSE RATIO OF 6.25-CM. SPHERE
(Compared to 25-cm. Sphere)

6.25-cm. diameter sphere gap, cm.	Voltage required to cause 6.25-cm. sphere to spark over at gaps greater than the diameter		Ratio impulse to 60 cycles, voltage
	60 cycles (max.)	Impulse measured by 25-cm. sphere (dia.) (max.)	
6	127	134	1.05
8	144	152	1.05
9	154	160	1.04
10	156	168	1.06
12	168	180	1.08
15	180	198	1.10

spark over a needle gap compared to the 60-cycle spark-over voltage. This indicates that the needle gap is very slow. A spark-over test for oil is also given at 60 cycles and for impulse. (See Table LXIX.) The impulse voltages show that oil requires greater energy than air to cause rupture.

Care must be taken when measuring the spark voltage on one gap by another gap in parallel; for instance, if a needle gap is set so as to just spark over when voltages of steep wave front or high frequency oscillation of a given value are applied, and a sphere gap is similarly set, and these two gaps are then placed in parallel, and the same impulse voltage applied, apparent discrepancy results. Spark-over will take place across one gap and not the other, even when the spacing on the non-sparking gap is decreased. This will be noticed in all cases where electrodes of different shapes are placed in parallel. When the impulse voltage is applied across the "fast" gap and the "slow" gap in multiple, the "fast" gap will spark over, relieving the stress, before the "slow" gap has time to act, even though the slow gap is set for a much lower voltage. The difference would not be apparent at commercial frequencies.

TABLE XXXVII.—IMPULSE RATIO OF NEEDLE GAP
(Impulse Circuit, Fig. 106)

Spacing of needle (cm.)	Voltage required to cause needle gap to spark over			Ratio impulse to 60 cycles, voltage
	60 cycles (max.)	Impulse measured by 25-cm. spheres (max.)	Impulse calculated from circuit constants (max.)	
2.30	27	32.5	32.5	1.20
3.19	35	45.5	46.0	1.30
3.95	41	55.0	56.0	1.35
4.44	45	65.0	64.0	1.40
5.34	50	73.0	74.0	1.45
10.00	76	133.0	135.0	1.75

An example of the impulse used by the author to obtain the data in Table XXXVII is shown in Fig. 106.[1] It reaches its maximum in 5×10^{-7} seconds and approximates a single half cycle of a 500,000-cycle wave. The calculated values of these voltages check approximately with those measured by the spheres. It thus appears that the sphere voltages are very little affected, even at this steep impulse.

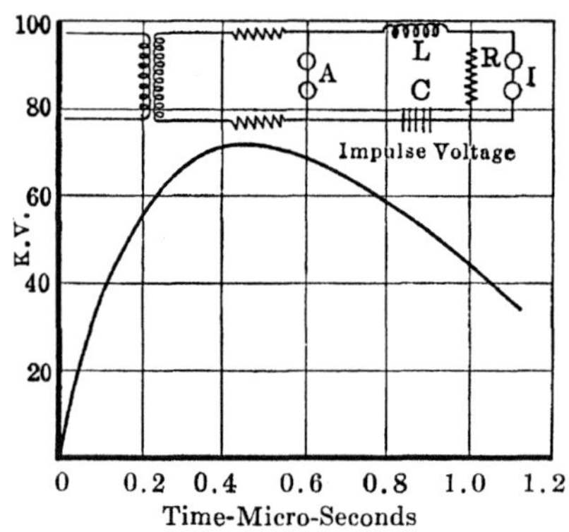

FIG. 106.—Impulse voltage and circuit producing it.[2]

The needle requires the maximum time of any gap as considerable air must be "ionized" before spark-over can result. The spheres are very little affected because corona does not precede spark-over, and the discharge is small and confined to a short path.

The line insulator is generally designed so that at commercial

[1] Impulse takes place at I due to discharge of condenser C, through R, L, and power arc at A. (See page 162.)

[2] One micro-second = one millionth of a second.

frequencies the spark-over voltage is lower than the puncture voltage. By applying a sufficient number of impulses at voltages higher than the 60-cycle puncture voltage, such an insulator may be punctured. During the time that the air is breaking down, this overstress produces small cracks in the porcelain. The effect is cumulative. The cracks gradually extend and puncture results. This subject is, hence, of great importance. In making spark measurements it is necessary to consider the spark lag, as otherwise totally erroneous conclusions may be reached. It is therefore necessary in making spark-over measurements on certain apparatus to guard against this. For instance, on line insulators there is generally considerable corona discharge long before spark-over. This discharge may cause oscillation which will affect the measuring gap and cause a lower voltage to arc over given distances than normal, or cause the gap to cause higher voltages than really exist. This is eliminated in practice by the use of high resistance in the measuring gap.

Another investigation of impulse voltages, indicating that time and energy are necessary to rupture insulation,[1] shows that "the disruptive discharge through a dielectric requires not merely a sufficiently high voltage, but requires a definite minimum amount of energy.

"The disruptive discharge does not occur instantly with the application, but a finite, though usually very small, time elapses after the application of the voltage before the discharge occurs. During this time the disruptive energy is supplied to the dielectric."

The disruptive energy of oil seems to be about 30 times greater than it is for air. This is further discussed in Chapter VI.

Effect of Altitude on the Spark-over Voltages of Bushings, Leads, and Insulators.—For non-uniform fields, as those around wires, spheres, insulators, etc., the spark-over voltage decreases at a lesser rate than the air density. The theoretical reasons for this have been given, as well as the laws for regular symmetrical electrodes, for cylinders and spheres.

It is, however, not possible to give an exact law covering all types of leads, insulators, etc., as every part of the surface has its effect. The following curves and tables give the actual test results on leads, insulators and bushings of the standard types. The correction factor for any other lead or insulator of the same

[1] Hayden and Steinmetz, A.I.E.E., June, 1910.

type may be estimated with sufficient accuracy. When there is doubt δ may be taken as the maximum correction. It will generally be advisable to take δ because the local corona point on leads and insulators will vary directly with δ. This is so because the corona must always start on an insulator in a field which is locally more or less uniform.

The tests were made by placing the leads or insulators in the large wooden cask, already referred to, exhausting the air to approximately $\delta = 0.5$, gradually admitting air and taking the spark-over voltage at various densities as the air pressure increased. The temperature was always read and varied between 16 and 25 deg. C.

At the start a number of tests were made to see if a spark-over in the cask had any effect upon the following spark-overs by ionization or otherwise. It was found that a number of spark-overs could be made in the cask with no appreciable effect. During the test, the air was always dry and the surfaces of the insulators were kept clean.

TABLE XXXVIII.—SUSPENSION INSULATOR

Bar. cm.	Vac. cm.	Pressure	Temp. cent.	δ	Kilovolts arc-over
75.4	37.4	38.0	22.0	0.50	121.0
75.4	34.3	41.1	22.0	0.54	131.0
75.4	30.0	45.4	22.0	0.60	144.0
75.4	26.4	49.0	22.0	0.65	158.5
75.4	23.0	52.4	22.0	0.70	165.0
75.4	19.3	56.0	22.0	0.74	177.5
75.4	17.5	57.9	22.0	0.87	183.2
75.4	15.0	60.4	22.0	0.80	195.0

TABLE XXXIX.—LEADS

Correction Factor for Leads Shown in Fig. 107

δ	(a)	(b)	(c)	(d)
1.00	1.00	1.00	1.00	1.00
0.90	0.92	0.91	0.92	0.92
0.80	0.83	0.82	0.83	0.85
0.70	0.74	0.72	0.75	0.77
0.60	0.70	0.65	0.64	0.66
0.50	0.61	0.56	0.54	0.57

TABLE XL.—POST AND PIN INSULATORS

Correction Factor for Insulators Shown in Fig. 108

δ	(a) Post	(b) Pin	(c) Pin
1.00	1.00	1.00	1.00
0.90	0.93	0.91	0.94
0.80	0.84	0.81	0.86
0.70	0.76	0.72	0.75
0.60	0.68	0.62	0.65
0.50	0.60	0.52	0.53

TABLE XLI.—SUSPENSION INSULATOR

Fig. 109

Correction Factor for Units in String as Follows

δ	Number of units				
	1	2	3	4	5
1.00	1.00	1.00	1.00	1.00	1.00
0.90	0.96	0.93	0.90	....	....
0.80	0.91	0.84	0.80	....	....
0.70	0.86	0.76	0.70	....	....
0.60	0.80	0.66	0.60	....	....
0.50	0.72	0.55	0.50	....	....

TABLE XLII.—SUSPENSION INSULATOR

Fig. 110

Correction Factor for Units in String as Follows

δ	Number of units				
	1	2	3	4	5
1.00	1.00	1.00	1.00	1.00	1.00
0.90	0.94	0.92	0.90	0.90	....
0.80	0.87	0.84	0.80	0.80	....
0.70	0.81	0.73	0.70	0.70	....
0.60	0.72	0.63	0.60	0.60	....
0.50	0.62	0.52	0.50	0.50	....

Table XXXVIII is a typical data sheet. Tables XXXIX–XLI give even values of δ and the corresponding measured correction factors. If the spark-over voltage is known at sea level

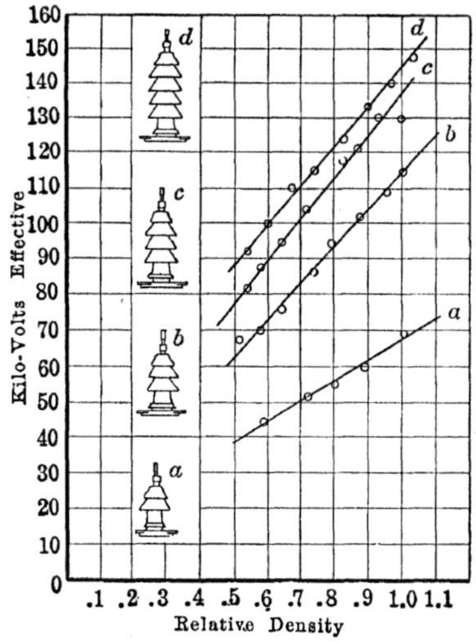

FIG. 107.—Variation of spark-over voltage of transformer leads with air density.

(*a*) 15.2 cm. high by 17.8 cm. dia.
(*b*) 21.6 cm. high by 17.8 cm. dia.
(*c*) 28 cm. high by 17.8 cm. dia
(*d*) 38.1 cm. high by 17.8 cm. dia.
Height measured from case to metal cap on top.

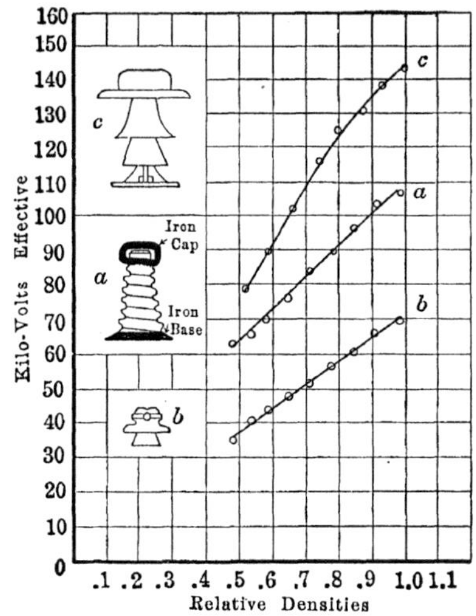

FIG. 108.—Variation of spark-over voltage insulators with air density.

(*a*) 30.2 cm. high.
(*b*) 13.5 cm. high by 16.8 cm. dia.
(*c*) 28.6 cm high by 36 cm. dia.

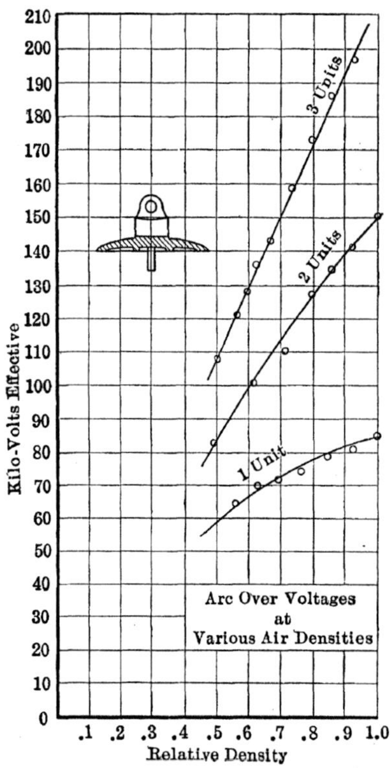

FIG. 109.—Suspension insulator. (Dia., 30 cm. Spacing 16.5 cm.)

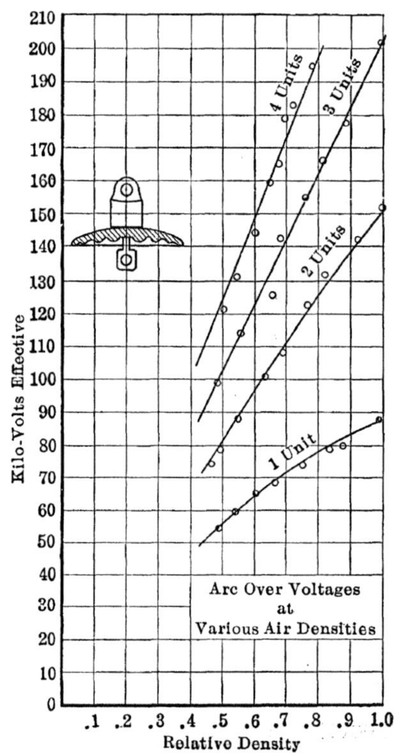

FIG. 110.—Suspension insulator. (Dia., 27 cm. Spacing 17 cm.)

or $\delta = 1$(76 cm. bar., temperature 25 deg. C.), the spark-over at any other value of δ may be found by multiplying by the corresponding correction factor. It will be noted that in most cases the correction factors are very nearly equal to δ. δ would be the correction factor in a uniform field and should be, as already stated, taken as such in most cases, especially where dirt and moisture enter, as in practice. Furthermore, it should be taken

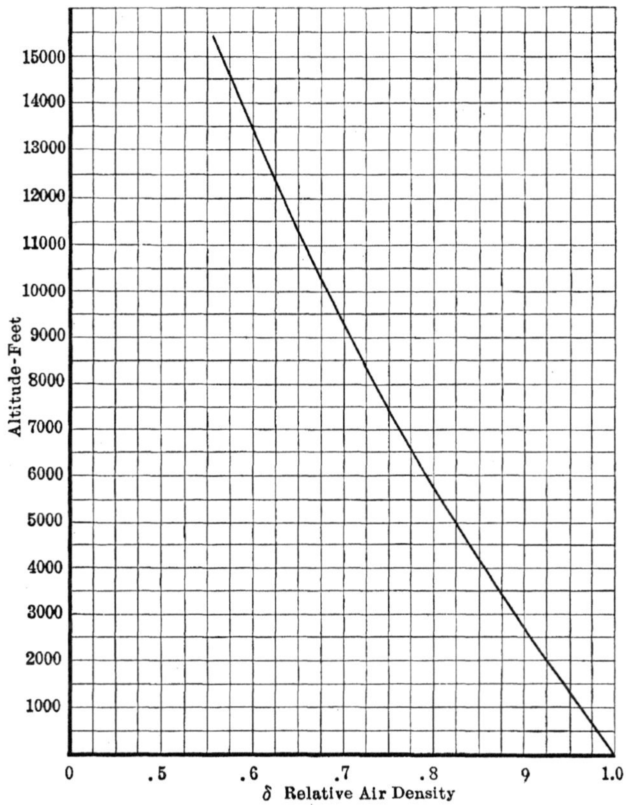

FIG. 111.—Approximate variation of air density with altitude.

as it is the actual correction factor for the starting point of local corona on insulators.

Fig. 111 is a curve giving different altitudes and corresponding δ at 25 deg. C. If the spark-over voltage is known at sea level at 25 deg. C., the spark-over voltage at any other altitude may be estimated by multiplying by the corresponding δ, or more closely if the design is the same as any in the tables, by the correction factor corresponding to δ. If the local corona starting point is known at sea level, it may be found for any altitude by multiplying by the corresponding δ. If the barometric pressure and temperature are known, δ may be calculated.

As an example of the method of making corrections: Assume a suspension insulator string of three units with a spark-over voltage of 205 kv. (at sea level 25 deg. C. temperature). $\delta = 1$. What is the spark-over voltage at 9000 ft. elevation and 25 deg. C.?

From Fig. 111, the δ corresponding to 9000 ft.

$$\delta = 0.71$$

Then the approximate spark-over voltage at 9000 ft., 25 deg. C. is

$$e_1 = 0.71 \times 205 = 145 \text{ kv.}$$

If this happens to be the insulator of Fig. 110, the correction factor corresponding to $\delta = 0.71$ is found in Table XLII, by interpolation to be 0.71. The actual spark-over voltage for the special case happens to check exactly with that given by δ. For practical work a correction may generally be made directly by use of Fig. 111.

The spark-over voltage of an insulator is 100 kv. at 70 cm. barometer and 20 deg. C. What is the approximate spark-over voltage at 50 cm. barometer and 10 deg. C.?

$$\delta_1 = \frac{3.92 \times 70}{273 + 30} = 0.94$$

$$\delta_2 = \frac{3.92 \times 50}{273 + 10} = 0.61$$

$$e_1 = 100 \times \frac{0.61}{0.94} = 65 \text{ kv.}$$

If the local corona starting point is known at sea level, it may be found very closely for any other altitude by multiplying by the correction δ.

CHAPTER V

CORONA LOSS

In the present chapter the corona loss is discussed. It has been thought worth while to go into details in the description of the apparatus, methods of making loss tests, and reducing data[1] as an example of an extremely large engineering investigation. Experimentally, the methods followed apply to any investigation; practically, many of the detailed observations have an important bearing; theoretically and experimentally the observed details are of importance and the methods of reducing data may be applied to other investigations.

Lines, Apparatus and Method of Test.—The Lines. The first investigation was made out of doors. The conductors used in

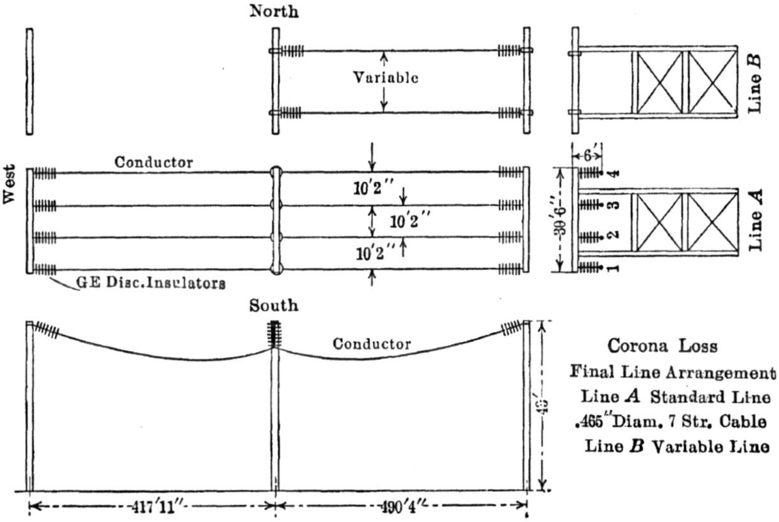

FIG. 112.—Experimental outdoor line.

this investigation were supported by metal towers arranged in two parallel lines of two spans each. The length of each span was approximately 0.150 km. These tower lines will be designated by *A* and *B* respectively. The conductors were strung in a horizontal plane with seven disk suspension insulators at each point

[1] Law of Corona, A. I. E. E., June, 1911.

of support. For preliminary tests four No. 3/0 B. & S. (1.18 cm. diameter) seven-strand, hard-drawn copper cables were put in place on each line. A seven-strand steel ground cable was also strung. After a number of tests had been made the ground cables were removed from line A. The conductors on line A, however, were kept in place as a standard throughout all the investigations. The conductors were removed from B, and the first span of this line was used to support various sizes of conductors at various spaces. (See Fig. 112.)

These lines were erected in a large field. The prevailing winds were from the west over open country, that is, free from smoke from the city and the factory on the east.

Test Apparatus.—A railroad track was run directly under the line, and the testing apparatus was housed in three box cars.

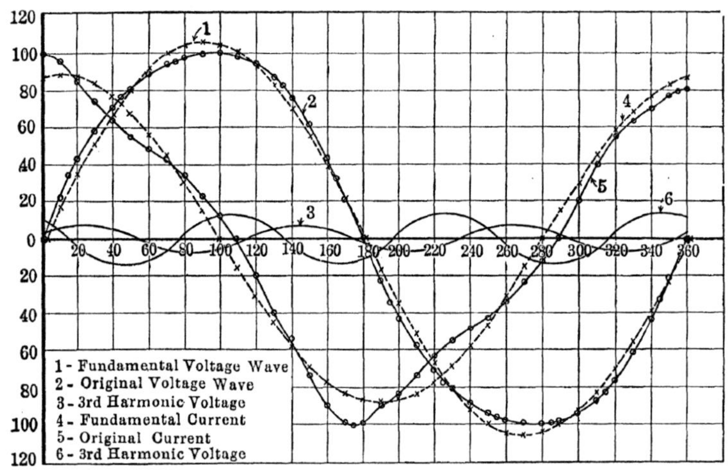

FIG. 113.—Analysis of applied voltage and corona current wave.

This proved a very convenient arrangement, as the cars could be quickly run back to the factory when changes or repairs were necessary.

Power was supplied from an old Thomson-Houston machine with a smooth core, pan cake winding on the armature; it gave a very good wave, and was used in all tests. See oscillogram and analyzed wave, Fig. 113. It was rated at 35 kw., but this rating was quite conservative. This alternator was belted to a d.c. motor.

The high voltage transformer and the testing apparatus were placed in car No. 2. The portion of the car roof over the transformer was made of heavy canvas. This could be quickly rolled back, and the leads from the line were dropped directly to the transformer terminal. By means of a framework and canvas cover the transformer could be protected from the weather, and investigations carried on during rain and snow storms. The power supply, speed and voltage, were all controlled from car No. 2. In fact all of the adjustments could be made from this car (see Fig. 114). The transformer was rated at 100 kw., 200,000 volts and 60 cycles. On the low side were four 500 volt coils. These coils could be connected in multiple or series for change of ratio. The high tension winding was opened at the neutral and taps were brought out for the ammeter, and current coil of the wattmeter. Three taps were also brought out here from the main winding for voltage measurement. (See Fig. 114.) The following tap ratios were thus obtained: 100/200,000; 200/200,000 and 200/200,000.

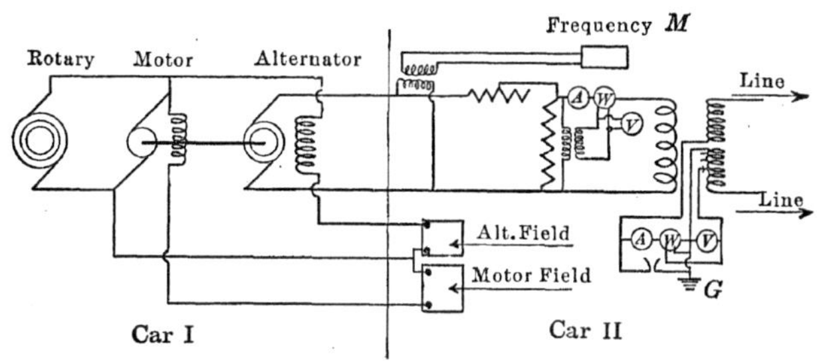

FIG. 114.—Circuit connections in corona loss measurements.

Car No. 3 served as a dark room for making photographs and visual tests on short wires and cables.

Methods of Test.—Accurate power measurements of corona are difficult to make, because of the nature of the load, low power factor and high voltage. It is not desirable to make the measurements on the low side because of the difficulty in separating the transformer iron and load losses, and these may be sometimes as large as the corona losses. In these tests the current coil of the wattmeter and the ammeter, were put in the high tension winding of the transformer at the neutral point, and the neutral was grounded. The voltage coil of the wattmeter was connected to a few turns of the high tension winding at the neutral.[1] All of the loss measurements were also duplicated on the low side as a

[1] A. B. Hendricks, A. I. E. E. Transactions, Feb., 1911.

TABLE XLIII.—EXPERIMENTAL LINE—A. CORONA LOSS—10–6–10. 4 P.M.

Low side total readings			High side total readings		
Volts	Amperes	Kilowatts	Kilovolts	Amperes	Kilowatts
		Line on			
395	16.5	0.40	80.5	0.077	0.10
435	17.9	0.60	90.5	0.087	0.13
490	20.5	0.70	101.6	0.101	0.17
535	22.6	0.80	111.1	0.112	0.22
590	24.7	1.00	120.4	0.119	0.28
635	27.1	1.10	130.2	0.135	0.35
680	29.2	1.40	139.2	0.145	0.45
735	31.6	1.80	150.0	0.158	0.68
780	33.4	2.40	159.0	0.169	1.10
812	35.3	3.30	165.8	0.178	1.80
830	36.3	3.60	169.0	0.181	2.40
862	37.5	5.12	176.6	0.190	3.60
893	39.2	6.35	183.2	0.199	4.70
914	40.5	7.50	187.0	0.207	6.00
975	43.9	11.40	200.0	0.227	9.30
1020	47.8	14.50	209.0	0.243	12.60
1050	49.3	17.00	214.2	0.253	14.60
1080	52.7	19.50	220.2	0.267	17.60
1125	55.5	22.80	223.4	0.283	20.30
		Line off			
400	1.14	0.40	80.0	0.005	0.05
500	1.37	0.62	100.5	0.007	0.10
600	1.63	0.82	121.5	0.008	0.15
718	1.87	1.18	143.5	0.010	0.21
812	2.05	1.45	161.0	0.011	0.28
905	2.29	1.80	181.0	0.013	0.35
1015	2.69	2.20	202.2	0.015	0.44
1095	3.25	3.64	217.0	0.017	0.54

Weather

Cloudy-rain in morning. Barometer 75 cm. Temperature: wet 10 deg. C., dry 12 deg. C.

Line and connections

(1 and 3) (2 and 4) ground wires in place.

Total conductor length............................ 109,500 cm.
Spacing.. 310 cm.
No. 3/0 seven-strand cable diameter................ 1.18 cm.
Transformer ratio.................................1000/200,000
Frequency.. 60 cycles.

check. Frequency was held at the test table by means of the motor field and a vibrating reed type of frequency meter.

Voltage was controlled in two ways—by the potentiometer method and by rheostats in the alternator field. By the potentiometer method is meant a resistance in series with the supply on the low side of the transformer for voltage control and a multiple resistance across the transformer, taking about three times the exciting current, to prevent wave distortion. When the leading current was very high a reactance was arranged to shunt the generator and approximately unity power factor could be held. This prevented overloading the generator and reduced wave shape distortion. For a set of tests at a given frequency the ratio of the main transformer was kept the same. Where losses at several frequencies were to be compared the main transformer ratio also was changed to keep the flux on the generator as nearly constant as possible—for instance, at 45 cycles a ratio of 500/200,000 would be used, while at 90 cycles a ratio of 1000 to 200,000 would be used. Wattmeters especially adapted to the tests were constructed. These were of the dynamometer type; each was provided with a 75-volt and 150-volt tap. The voltmeter coil ratio on the transformer and the wattmeter tap were always changed to give the best reading. Four wattmeters were used in these tests. The meters were all carefully calibrated in the laboratory at unity power factor and at 0.10 leading power factor, at both 25 and 60 cycles.

Humidity, temperature and barometric pressures, as well as general weather observations, were taken during each test.

Indoor Line.—Later an extensive investigation was made in a large laboratory room, 17 meters wide by 21 meters long. The lines were strung diagonally between movable wooden towers. Strips of treated wood 1.25 cm. square by 80 cm. long were used as insulators. The total length of conductor possible with four wires was 80 meters. By this arrangement it was possible to make a more complete study on the smaller sizes of conductors, and also to extend the investigation over a greater frequency range. The apparatus used was otherwise the same as in the outdoor tests.

The Quadratic Law.—Table XLIII is a typical data sheet for Line A. Fig. 115 and Fig. 118 show the characteristic corona curves. The corrected values for Table XLIII are recorded in Table XLIV.

TABLE XLIV.—CORONA LOSS, OBSERVED VALUES CORRECTED FROM TABLE XLIII

Kilovolts between lines e^1	Line ampere	Kilovolts to neutral e	Kilowatts line loss p	K.v.a.	Power factor
80.5	0.072	40.2	....	5.80	....
90.5	0.081	45.2	....	7.33	....
101.6	0.094	50.8	....	9.55	....
111.1	0.104	55.5	....	11.55	....
120.4	0.111	60.2	0.11	13.40	0.008
130.2	0.126	65.1	0.15	16.40	0.009
139.2	0.135	69.6	0.22	18.80	0.012
150.0	0.147	75.0	0.40	22.10	0.018
159.0	0.157	79.5	0.79	25.00	0.032
165.8	0.166	82.9	1.42	27.60	0.051
169.0	0.168	84.5	2.04	28.40	0.072
176.6	0.177	88.3	3.21	31.20	0.103
183.2	0.185	91.6	4.28	33.90	0.126
187.0	0.193	93.5	5.55	36.10	0.154
200.0	0.212	100.0	8.78	42.40	0.207
209.0	0.227	104.5	12.02	47.50	0.253
214.2	0.237	107.1	13.99	50.70	0.276
220.2	0.250	110.1	16.94	55.10	0.307

The shape of the curve between kilovolts and kilowatts suggests a parabola. After trial it was found that the losses above the knee of the curve follow a quadratic law. Below the knee it was found that the curve deviates from the quadratic law. This variation near the critical voltage is due to dirt spots, irregularities and other causes as discussed later. The main part of the curve may be expressed by[1]

$$p = c^2(e - e_o)^2 \qquad (32)$$

where

p = the line loss.
e = kilovolts to neutral.
e_o is called the disruptive critical voltage, measured in kilovolts to neutral.

The meaning of e_o and c^2 will be considered later. The best mechanism of evaluation of constants for a given set of tests may now be considered. Equation (32) may be written

$$\sqrt{p} = c(e - e_o)$$

[1] F. W. Peek, Jr., Law of Corona, A. I. E. E., June, 1911.

then if the quadratic law holds, the curve between $\sqrt{p}$ and e will be a straight line. e_o will be the point where the line cuts the e axis, and c will be the slope of the line (see Fig. 116); e_o and c may be evaluated graphically in this way. It is difficult to know how to draw the line accurately and give each point the proper weight. To do this the $\Sigma\Delta$ method is used, as follows:[1]

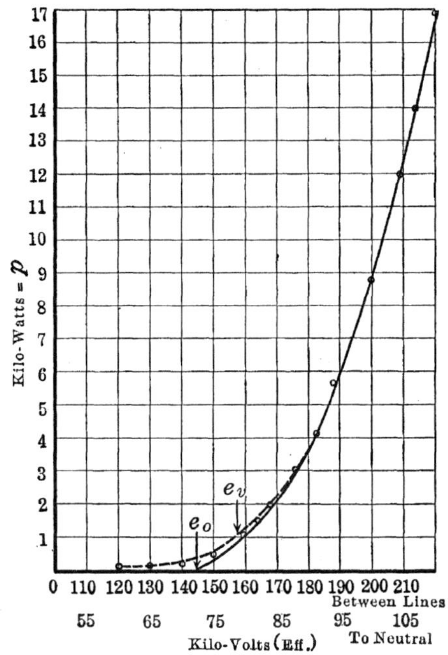

FIG. 115.—Characteristic corona loss curve for large stranded conductor.

Line A conductors 1–2–3–4. 3/0, 7-strand cable, diameter, 1.18 cm. Total conductor length, 109,500 cm. Spacing, 310 cm. Points, measured values. Curve calculated from $p = 0.0115\ (e - 72.1)^2$. e_o = Disruptive critical voltage. e_v = Visual critical voltage. Test table XLIV.

The values of e and p for the set of readings to be investigated are first tabulated and a curve plotted (Fig. 116). All points that differ greatly from the straight line are eliminated as probably in error, or, as at the lower part of the curve, following a different law. The remaining readings are taken and formed into two groups, each of an equal number of readings.

$$\begin{array}{lll} \text{Group 1.} & \Sigma_1 e & \Sigma_1\sqrt{p} \\ \text{Group 2.} & \Sigma_2 e & \Sigma_2\sqrt{p} \end{array}$$

[1] Steinmetz, Engineering Mathematics, page 232.

Then
$$\Delta\Sigma e = \Sigma_1 e - \Sigma_2 e \qquad \Delta\Sigma\sqrt{p} = \Sigma_1\sqrt{p} - \Sigma_2\sqrt{p}$$
$$\Sigma\Sigma e = \Sigma_1 e + \Sigma_2 e \qquad \Sigma\Sigma\sqrt{p} = \Sigma_1\sqrt{p} + \Sigma_2\sqrt{p}$$
$$c = \frac{\Delta\Sigma\sqrt{p}}{\Delta\Sigma e}$$
$$e_o = \frac{\Sigma\Sigma e - \frac{\Sigma\Sigma\sqrt{p}}{c}}{n}$$

where n is the number of points used.

Thus e_o and c are determined.

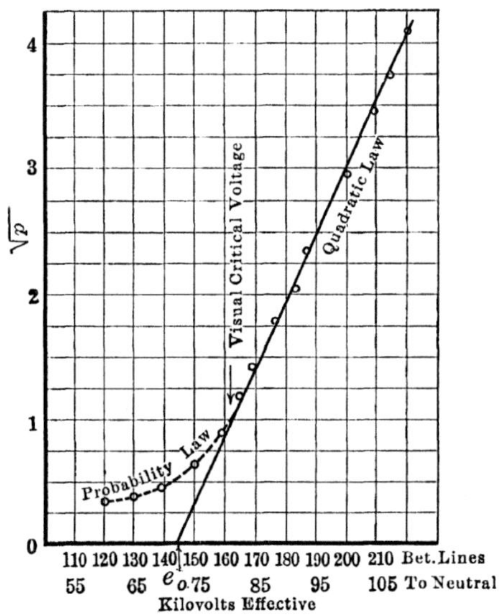

FIG. 116.—Corona loss curve plotted between $\sqrt{p}$ and e. (Large conductor. Data same as Fig. 115.)

TABLE XLV.—CORONA LOSS, CALCULATED VALUES FOR FIG. 115.

$$p = c^2(e - e_o)^2$$
$$p = 0.0115\ (e - 72.1)^2$$

Kilovolts between lines e'	Kilovolts to neutral e	Kilowatts $p = c^2(e - e_o)^2$	Kilovolts between lines e'	Kilovolts to neutral e	Kilowatts $p = c^2(e - e_o)^2$
144.2	71.2	0.0	183.2	91.6	4.17
150.0	75.0	0.10	187.0	93.5	5.08
159.0	79.5	0.63	200.0	100.0	9.03
165.8	82.9	1.34	209.0	104.5	12.10
169.0	84.5	1.77	214.2	107.1	14.10
176.6	88.3	3.02	220.2	110.1	16.70

TABLE XLVI.—CORONA LOSS

Method of Reducing (Data from Table XLIV)

Kilovolts between line e'	Kilovolts to neutral e	Kw., p	$\sqrt{p}$
120.4	60.2	0.11	0.332
130.2	65.1	0.15	0.388
139.2	69.6	0.22	0.470
150.0	75.0	0.40	0.632
159.0	79.5	0.79	0.889
165.8	82.9	1.42	1.192
169.0	84.5	2.04	1.428
176.6	88.3	3.21	1.792
183.2	91.6	4.28	2.069
187.0	93.5	5.55	2.356
200.0	100.0	8.78	2.963
209.0	104.5	12.02	3.467
214.2	107.1	13.99	3.740
220.2	110.1	16.94	4.116

Total conductor length . . . . 109,500 cm.
Spacing . . . . 310 cm.
No. 3/0 seven-strand cable diameter . . . . 1.18 cm.

e	$\sqrt{p}$
91.6	2.069
93.5	2.356
100.0	2.963
104.2	3.467
107.1	3.740
110.1	4.116

$\Sigma_2 e = 285.1$

$\Sigma_1 e = 321.4$

$\Delta\Sigma e = 36.6 \qquad \Delta\Sigma\sqrt{p} = 3.935$

$\Sigma\Sigma e = 606.8 \qquad \Sigma\Sigma\sqrt{p} = 18.711$

$$c = \frac{\Delta\Sigma\sqrt{p}}{\Delta\Sigma e} = 0.107$$

$$c^2 = 0.0115$$

$\Sigma_2\sqrt{p} = 7.388$

$\Sigma_1\sqrt{p} = 11.323$

$e_o = 72.1$

$e_o' = 144.2$

$$e_o = \frac{\Sigma\Sigma e - \dfrac{\Sigma\Sigma\sqrt{p}}{c}}{n}$$

Table XLVI shows the method of reducing. The curve, Fig. 115, is drawn from the equation $p = 0.0115\,(e - 72.1)^2$. The circles show the experimental values, and indicate where the losses deviate from the quadratic law.

Table XLVII gives a similar set of data for a small wire. The results are plotted in Figs. 117 and 118.

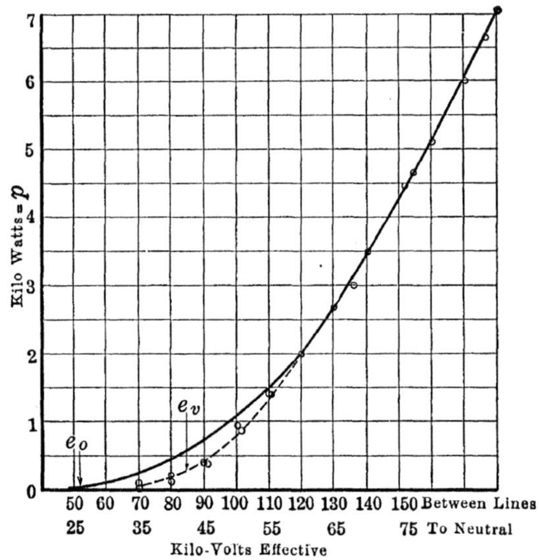

FIG. 117.—Characteristic corona loss curve for small wire. No 8 copper wire. Diameter, 0.328 cm. Total length, 29,050 cm. Spacing, 0.328 cm. Table XLVII, Line B.

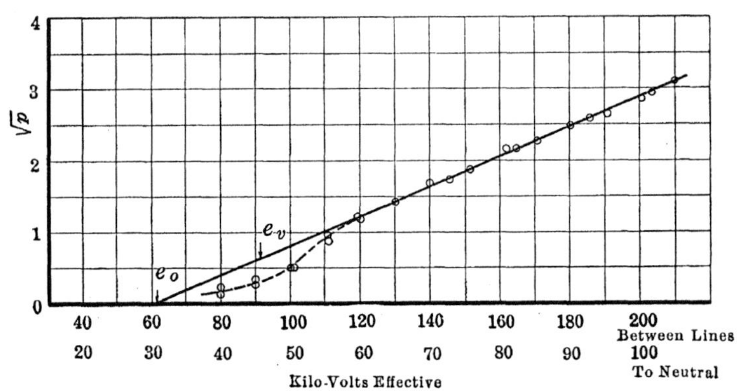

FIG. 118.—Corona loss curve plotted between $\sqrt{p}$ and e. (Small conductor. Data same as Fig. 117.)

TABLE XLVII.—CORONA LOSS

$\Sigma\Delta$ Method of Reducing

Kilovolts between line e'	Kilovolts to neutral e	Kilowatt line loss p	$\sqrt{p}$
70.0	35.0	0.02	0.14
80.0	40.0	0.07	0.26
91.2	45.6	0.26	0.51
101.3	50.6	0.85	0.92
110.0	55.0	1.42	1.19
120.0	60.0	2.02	1.42
130.0	65.0	2.71	1.65
141.5	70.7	3.51	1.87
70.0	35.0	0.06	0.24
80.0	40.0	0.10	0.32
90.5	45.2	0.26	0.51
101.3	50.6	0.96	0.98
109.9	54.9	1.43	1.20
152.0	76.0	4.45	2.11
160.4	80.2	5.17	2.27
170.0	85.0	6.06	2.46
180.6	90.3	7.04	2.65
190.6	95.3	8.26	2.87
200.0	100.0	9.52	3.08
193.6	96.8	8.60	2.93
176.0	88.0	6.65	2.58
155.0	77.5	4.66	2.16
136.0	68.0	3.01	1.73

Total conductor length.. 29,050 cm.
Spacing.. 183 cm.
No. 8 H. D. copper wire—diameter.......................... 0.328 cm.
Temperature.. 1.5
Barometer... 76.6

e	$\sqrt{p}$	e	$\sqrt{p}$
100.0	3.08	80.2	2.27
96.8	2.93	77.5	2.15
95.3	2.87	68.0	1.73
90.3	2.65	60.0	1.42
85.0	2.46	88.0	2.58
467.4	13.99	373.7	10.16

$\Sigma\,\Sigma e = 841.1 \qquad \Delta\Sigma e = 93.7$

$\Sigma\,\Sigma\sqrt{p} = 24.15 \qquad \Delta\Sigma\sqrt{p} = 3.80$

$$e_0 = \frac{841.1 - \dfrac{24.15}{0.0406}}{10}$$

$$c = \frac{3.80}{93.7} = 0.0408$$

$$c^2 = 0.00164$$

$$p = 0.00164(e - 24.6)^2$$

To further investigate the law it is now necessary to determine the various factors affecting e_o and c^2. These will be taken up

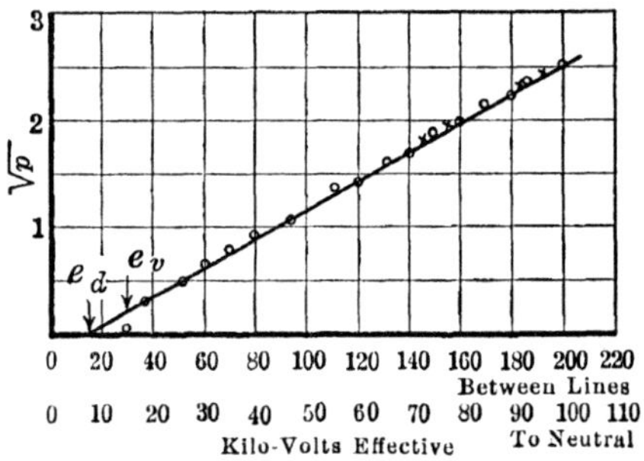

FIG. 119.—Corona loss plotted between $\sqrt{p}$ and e to illustrate the quadratic law.

Phosphor-bronze conductor. Diameter, 0.051 cm. Spacing, 366 cm. Total length, 29,050 cm. Temp.,−6.5° C. Bar., 77.3 cm. Line B.

under separate headings. The loss near the critical point will then be discussed.

In Fig. 119, $\sqrt{p}$ and e are plotted. This is an especially interesting curve on account of its range. The measurements are

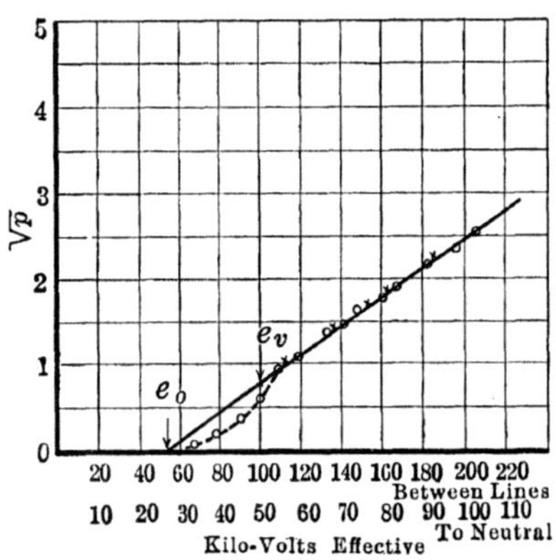

FIG. 120.

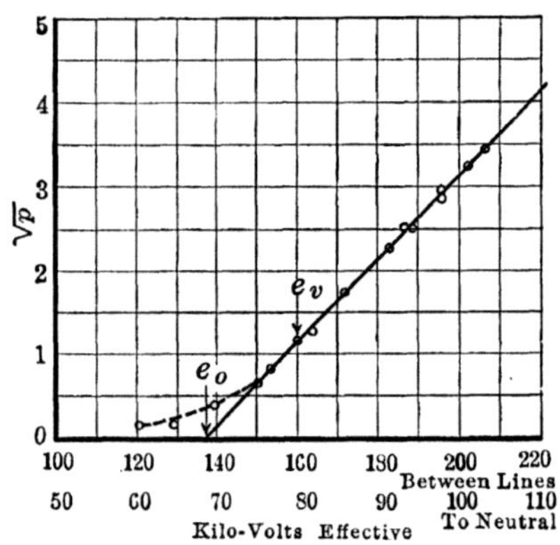

FIG. 121.

Corona loss plotted between $\sqrt{p}$ and e to illustrate the quadratic law.

FIG. 120.—No 8 copper wire. Diameter, 328 cm. Total length, 29,050 cm. Spacing, 244 cm. Temp. 1.5. Bar., 76.6. Line B.

FIG. 121.—3/o, 7-strand weathered cable. Diameter, 1.18 cm. Spacing, 310 cm. Total length, 109,500 cm. Temp., 16. Bar., 75.3. Line A.

taken up to 20 times the disruptive critical voltage, and show how well the quadratic law holds. Figs. 120 and 121 are plotted in the same way to illustrate the quadratic law.

Frequency.—To determine the way that frequency enters into the power equation

$$p = c^2(e - e_o)^2$$

a series of loss curves were taken on line A at various frequencies.

These tests indicated that the loss varied almost directly with the frequency over the range investigated. The data are given in Tables XLVIII and XLIX. Thus when the frequency does not

TABLE XLVIII.—LINE A. CONDUCTOR 2–3. TOTAL LENGTH 54,750 CM.

Kilovolts between lines	210	200	190	180
f	Kilowatts loss			
47	4.9	3.2	2.0	1.3
60	5.1	3.7	2.6	1.6
70	6.1	4.3	2.8	1.7
80	6.9	5.0	3.4	1.9
....	....	....	3.6	2.1
90	7.5	5.6	3.8	2.6
100	8.0	5.6	3.7	2.4
115	9.3	7.1	5.0	3.4

TABLE XLIX.—LINE A. CONDUCTORS 1–2–3–4. TOTAL LENGTH 109,500 CM.

Kilovolts between lines	210	200	190	180
f	Kilowatts loss			
50	9.24	6.30	3.80	2.20
60	10.50	7.30	4.65	2.65
70	12.1	8.60	5.50	3.20
80	14.0	10.20	6.80	3.95

vary greatly from 60 cycles the equation may be written

$$p = af(e - e_o)^2$$

It was then decided to make the investigation over a greater range of frequency, and on shorter lines to prevent wave distortion due to load, etc. These measurements were made on the indoor line. Table L gives data for a 0.333-cm. radius wire. Care was taken to keep the wave shape as nearly constant as possible. The results are plotted in Fig. 122. e_o is the same for all fre-

TABLE L.—CORONA LOSS AT DIFFERENT FREQUENCIES

New Galvanized Cable

Radius = 0.334 cm. Total length = 0.0815 km.
Spacing = 61 cm. δ = 1.00
Effective kv. Indoor line

Kilovolts to neutral e_n	Kilowatts per km. (p)	$\sqrt{p}$	Kilovolts to neutral e_n	Kilowatts per km. (p)	$\sqrt{p}$	Kilovolts to neutral e_n	Kilowatts per km. (p)	$\sqrt{p}$	Kilovolts to neutral e_n	Kilowatts per km. (p)	$\sqrt{p}$
Test 50*B*, 40 cycles			Test 51*B*, 60 cycles			Test 52*B*, 90 cycles			Test 53*B*, 120 cycles		
50.5	1.10	1.05	54.0	3.31	1.82	89.5	61.5	7.84	45.0	0.67	0.82
53.5	2.57	1.59	62.5	9.15	3.02	85.8	54.0	7.35	50.7	2.88	1.69
56.2	4.15	2.04	68.0	14.55	3.88	87.0	56.0	7.48	53.7	6.73	2.59
59.2	5.84	2.42	72.0	20.10	4.48	82.8	48.2	6.93	56.2	10.30	3.21
64.0	9.10	3.02	78.0	28.45	5.20	77.5	37.50	6.13	60.2	15.60	3.94
66.8	11.55	3.40	87.3	40.20	6.32	74.0	31.60	5.64	69.0	30.00	5.46
71.5	15.95	3.99	91.5	49.70	7.02	66.0	18.90	4.35	72.7	37.40	6.10
73.7	18.44	4.30	82.0	34.60	5.83	60.4	12.75	3.75	71.0	32.00	5.65
70.7	15.23	3.95	72.8	21.10	4.67	50.3	4.90	1.40	75.5	41.30	6.42
67.5	12.30	3.50	52.7	5.23	2.90	54.3	5.02	2.24	81.0	53.50	7.30
65.0	10.20	3.36	56.0	5.65	2.32	58.4	9.45	3.08	87.5	69.80	8.35
60.2	6.64	2.58	61.5	10.20	3.20	......	......	......	95.0	90.00	9.50
58.2	5.28	2.29	67.2	15.75	3.90	......	......	......	80.0	51.20	7.15
46.7	0.80	0.89	70.2	19.30	4.38	......	......	......	75.5	41.50	6.42
......	......	......	......	......	......	......	......	......	70.5	29.70	5.42
......	......	......	......	......	......	......	......	......	66.5	25.60	5.06
......	......	......	......	......	......	......	......	......	59.5	13.65	3.69

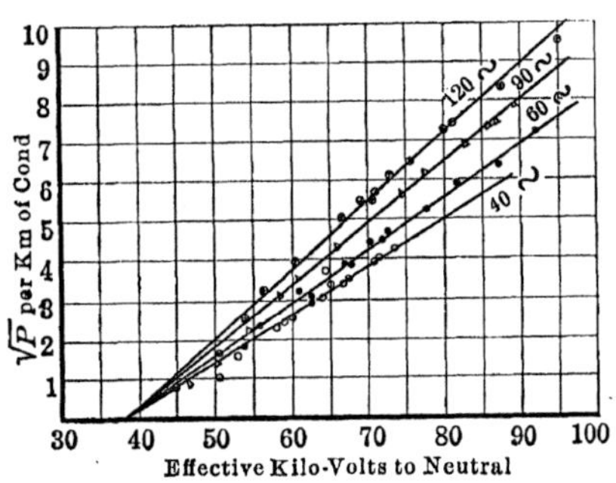

FIG. 122.—Corona loss curves at different frequencies. (Plotted from Table L.)

quencies but the slope c varies with the frequency. Values of c^2 for various frequencies from 30 to 120 cycles and three different sizes of conductor are given in Table LI, and plotted in Fig. 123.

TABLE LI.—VARIATION OF c^2 WITH FREQUENCY
(Indoor Line)

Frequency	Tests 1–6B, $r = 0.032$ cm., $s = 61$ cm., Length = 0.0815 km., $\delta = 1.00$		Tests 37–40B, $r = 0.105$ cm., $s = 61$ cm., Length = 0.0815, $\delta = 1.00$		Tests 50–53B, $r = 0.334$ cable, $s = 61$ cm., Length = 0.0815	
	c^2 per km.	e_o	c^2 per km.	e_o	c^2 per km.	e_o
30	0.0052	9.5	0.0071	21.5	....	....
40	0.0061	9.5	0.0092	21.5	0.0139	38.0
60	0.0078	9.5	0.0119	20.5	0.0173	38.0
75	0.0092	9.5	0.0144	20.5	....	38.0
90	0.0107	9.5	0.0157	20.5	0.0240	38.0
120	0.0134	9.5	0.0203	19.0	0.0297	38.0

The points over this measured range lie on a straight line, and this line extended cuts the frequency axis at -25. This seems

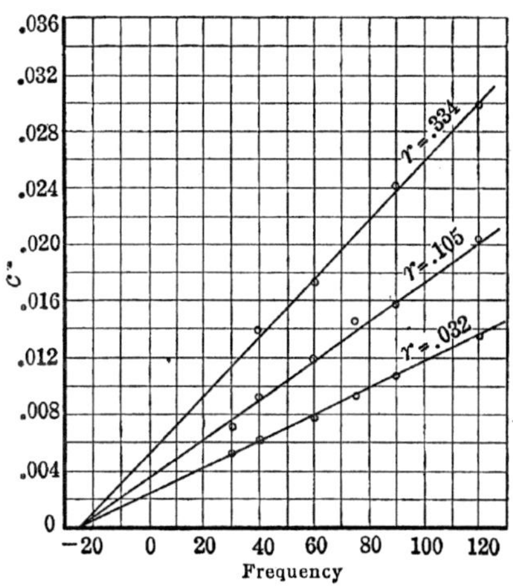

FIG. 123.—Variation of c^2 with frequency. (Data from Table LI.)

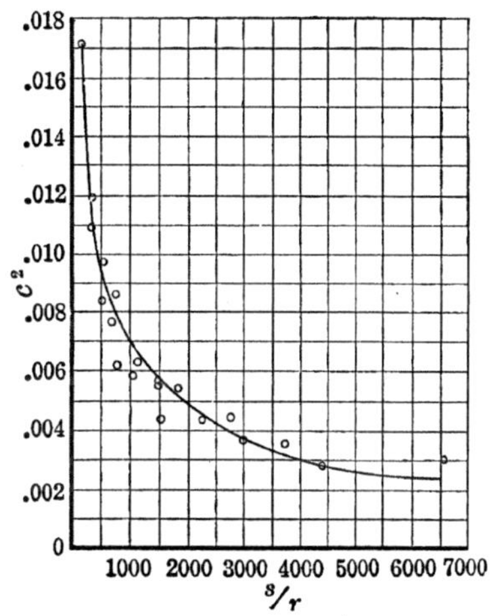

FIG. 124.—Variation of c^2 with s/r.

to mean for a given wire and spacing a constant loss plus a loss which varies directly as the frequency.

The equation may be written

$$p = a(f + 25)(e - e_o)^2$$

At zero frequency the equation reduces to

$$p = 25a(e - e_o)^2$$

This is not necessarily the direct-current loss but is probably lower as the maximum voltage is applied for less time. Watson has made laboratory measurements of d.c. loss, using an influence machine as the source of power.[1] Some of these measurements are compared with the a.c. loss in Table LII, for the same maximum voltages; the difference between the d.c. loss and the a.c. loss for the same effective voltages is *much* greater.

TABLE LII.—COMPARISON OF D.C. CORONA LOSS WITH LOSS AT 60 CYCLES FOR SAME MAXIMUM VOLTAGE

Voltage gradient g	D.c. amps. i	D.c. kilovolts e	D.c. loss p kw./km.	Corresponding a.c. effective kilovolts e_1	Measured 60-cycle loss p_1 kw./km.	Ratio $\frac{\text{60 cycles}}{\text{d.c. loss}}$ $\frac{p_1}{p}$
61	0.003	55	0.16	39.8	0.25	1.52
66	0.004	59	0.24	42.0	0.37	1.58
69	0.005	62	0.31	44.0	0.50	1.61
75	0.007	67	0.47	48.0	0.80	1.67

$s = 100$ cm. $r = 0.0597$ cm.

D.c. measurements from Watson, Institute of Electrical Engineers.

Relation between c² and s/r.—The power equation may be written over the commercial range of frequency

$$p = c^2(e - e_o)^2 = a(f + 25)(e - e_o)^2$$

where the relation between c^2 and s/r has not to this point been investigated. The relation will first be determined for the practical sizes of conductors at practical spacings of the outdoor line, and later over greater range from the indoor data. Only 60-cycle values will be used, as the data at this frequency are very complete. In Table LIII are values of c^2 for various sizes of wire and cable at various spacings. c^2 varies greatly with the radius of the conductor r and the spacing s. Plotting s/r and c^2 a curve is obtained that suggests a hyperbola. The curve between log c^2 and log s/r is a straight line. Therefore the following relation between c^2 and s/r is established.

$$c^2 = A\ (s/r)^d \tag{33}$$

The fair weather value of c^2 for standard line A 1–2–3–4 may now be examined in Table LIV.

[1] Watson, Journal Institution of Electrical Engineers, June, 1910.

TABLE LIII.—EXPERIMENTAL VALUES—LINES A AND B

Relation between c^2 and s/r

Test No.	Diameter, cm.	$\frac{s}{r}$	δ	$c^2 \times 10^5$ per km.	Style of conductor	Material
95	0.168	6550	1.07	280	Wire	Gal. iron
92	0.168	4880	1.07	256	Wire	Gal. iron
86	0.168	3700	1.08	326	Wire	Gal. iron
138	0.328	2980	1.10	331	Wire	Copper
94	0.168	2730	1.07	410	Wire	Gal. iron
137	0.328	2230	1.10	391	Wire	Copper
91	0.168	1820	1.07	506	Wire	Gal. iron
128	0.518	1530	1.05	412	Wire	Copper
136	0.328	1490	1.10	513	Wire	Copper
82	0.585	1480	1.07	513	Cable	Gal. iron
135	0.328	1120	1.10	570	Wire	Copper
94*a*	0.168	1090	1.07	633	Wire	Gal. iron
77	0.585	1060	1.07	543	Cable	Gal. iron
79	0.585	770	1.08	571	Cable	Gal. iron
134	0.328	740	1.10	772	Wire	Copper
126	0.518	700	1.11	687	Wire	Copper
18	1.181	525	1.02	950	Cable	Copper
80	0.585	520	1.07	784	Cable	Gal. iron
125	0.518	350	1.11	1070	Wire	Copper
73	0.585	310	1.07	1018	Cable	Gal. iron
100	0.953	193	1.08	1584	Cable	Gal. iron

TABLE LIV.—RELATION OF c^2 TO $\frac{1}{\delta}$ FOR MAIN EXPERIMENTAL LINES

Line A 1–2–3–4

(Standard Line A)

Test No.	$c^2 \times 10^5$ per km.	$\frac{1}{\delta}$	δ	Test No.	$c^2 \times 10^5$ per km.	$\frac{1}{\delta}$	δ
18	945	0.982	1.020	103	980	0.901	1.112
36	1050	0.966	1.037	104	892	0.878	1.138
37	945	0.959	1.043	105	890	0.866	1.158
84	945	0.933	1.074	109	869	0.928	1.078
101	955	0.925	1.081	119	991	0.888	1.127

It is seen that these values are not exactly constant but apparently vary with the temperature and barometric pressure. c^2 and $1/\delta$ in curve, Fig. 125, suggest that c^2 varies as $1/\delta$. Multiplying by δ then, reduces c^2 to the standard temperature of 25

TABLE LV

(Corrected c^2 and s/r)

Test No.	Diameter, cm.	s/r	$c^2 \times 10^5$ read per km.	$c^2 10^5$ corrected to 25° C., 76 cm. bar.	Corr. factor δ	$\log(c^2 10^5)$ corrected to 25° C., 76 cm. bar.	$\log \frac{s}{r}$	Style of conductor
95	0.168	6550	280	300	1.07	5.704	8.787	Wire
92	0.168	4880	256	275	1.07	5.620	8.492	Wire
86	0.168	3700	326	347	1.08	5.852	8.215	Wire
138	0.328	2980	331	364	1.09	5.892	8.000	Wire
137	0.328	2230	391	429	1.09	6.062	7.709	Wire
77	0.585	1060	543	582	1.07	6.372	6.966	Cable
126	0.518	700	687	755	1.10	6.637	6.551	Wire
18	1.181	525	950	965	1.02	6.821	6.263	Cable
80	0.585	520	784	840	1.07	6.733	6.254	Cable
125	0.518	350	1070	1190	1.11	7.083	5.858	Wire
73	0.585	310	1018	1090	1.07	6.995	5.763	Cable
100	0.953	193	1584	1710	1.08	7.405	5.263	Cable

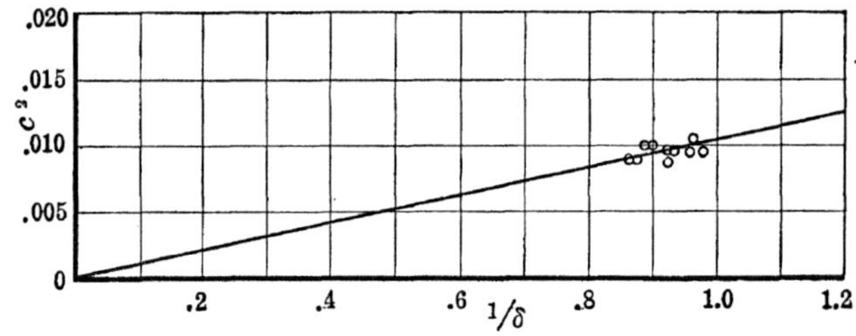

FIG. 125.—Relation between c^2 and $1/\delta$.

TABLE LVI.—ΣΔ REDUCTION OF RELATION OF c^2 TO $\frac{s}{r}$

(c^2 Corrected to 25 deg. C.—76 cm. Barometric Pressure)

Test No.	$\log_\epsilon (c^2 10^5)$	$\log_\epsilon \frac{s}{r}$	Test No.	$\log_\epsilon (c^2 10^5)$	$\log_\epsilon \frac{s}{r}$
95	5.704	8.787	126	6.637	6.551
92	5.620	8.492	18	6.871	6.263
86	5.852	8.215	80	6.733	6.254
138	5.892	8.000	125	7.083	5.858
137	6.062	7.709	73	6.995	5.736
77	6.372	6.966	100	7.444	5.263
	35.502	48.169		41.763	35.925

$$\Sigma_1 \log_\epsilon (c^2 10^5) = 35.50 \qquad \Sigma_1 \log_\epsilon \frac{s}{r} = 48.17$$

$$\Sigma_2 \log_\epsilon (c^2 10^5) = 41.76 \qquad \Sigma_2 \log_\epsilon \frac{s}{r} = 35.92$$

$$\Delta\Sigma \log_\epsilon (c^2 10^5) = -6.26 \qquad \Delta\Sigma \log_\epsilon \frac{s}{r} = 12.25$$

$$\Sigma\Sigma \log_\epsilon (c^2 10^5) = 77.26 \qquad \Sigma\Sigma \log_\epsilon \frac{s}{r} = 84.09$$

$$d = \frac{\Delta\Sigma \log_\epsilon c^2 \times 10^5}{\Delta\Sigma \log_\epsilon \frac{s}{r}} = \frac{-6.26}{12.25} = -0.51 \cong -0.50$$

$$c^1 = \frac{\log (c^2 10^5) - d\Sigma\Sigma \log_\epsilon \frac{s}{r}}{n} = 9.92$$

$$\log (c^2 10^5) = -0.5 \log_\epsilon \frac{s}{r} + 9.92$$

$$c^2 = 20{,}500 \sqrt{\frac{r}{s}} 10^{-5}$$

deg. C. and 76 cm. barometric pressure. This particular correction is not satisfactory as the range of δ is small. It seems the best until more complete data are obtained. In Table LV all values of c are corrected to 25 deg. C. and 76 cm. barometer, and the constants calculated by the $\Sigma\Delta$ method in Table LVI. This gives: $c^2 = 20{,}500\sqrt{r/s} \times 10^{-5}$ per km. of total conductor at 25 deg. C., 76 cm. barometer and 60 cycles.

Curve 124 is plotted from the points calculated in Table LVII, while the circles show the actual experimental points.

TABLE LVII.—CALCULATION OF CURVE No. 124 FROM $\left(c^2 \times 10^5 = 20{,}500\sqrt{\frac{r}{s}}\right)$

$\frac{s}{r}$	$\sqrt{\frac{r}{s}}$	$c^2 \times 10^5$	$\frac{s}{r}$	$\sqrt{\frac{r}{s}}$	$c^2 \times 10^5$
250	0.0633	129	2000	0.0224	460
500	0.0447	915	3000	0.0183	376
1000	0.0316	648	4000	0.0158	324
1500	0.0258	530	6000	0.0128	262

Curve 126 shows a straight-line relation between log s/r and log c^2. The equation for the power loss at 25 deg. C. and 76 cm. barometric pressure and any frequency may now be written

$$p = 241(f + 25)\sqrt{r/s}(e - e_o)^2 10^{-5} \qquad (34)$$

where p = the energy loss per kilometer of conductor in kilowatts.

e = kilovolts to neutral.

e_o = disruptive critical kilovolts to neutral at 25 deg. C. and 76 cm. barometric pressure.

f = the frequency in cycles per second.

r = the radius of the conductor in cm.

s = the distance between conductor centers in cm.

The value of e_o varies with the radius of the conductor r, and the spacing s, and will be discussed later.

Relation between c², and r and s for Small Conductors and Small Spacings. (Indoor Line).—The loss was investigated for very small conductors at large and small spacings, and for large conductors at small spacings, on the indoor line. The conductors ranged from 0.025 cm. to 0.46 cm. in radius, and the

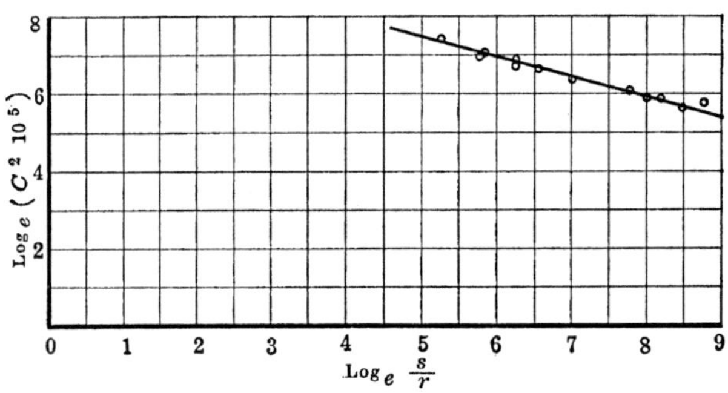

FIG. 126.—Determination of equation between c^2 and s/r.

spacings from 12.5 to 275 cm. This investigation is, hence, an extension of the above investigation beyond the practical range. The quadratic law still holds—the relation between c^2, r and s, however, becomes more complicated. This is also true of the disruptive critical voltage. Below a spacing of about 15 cm. it is difficult to express the loss in terms of a law, as the results are erratic, probably due to the great distortion of the field which is augmented when the corona starts, greatly increasing the loss above the quadratic.

For conductors 0.025 cm. in radius and above, and 15 cm. spacing and above, this data shows that

$$c^2 = 20{,}500\sqrt{\frac{r + \frac{6}{s} + 0.04}{s}}\,10^{-5} \text{ per km. of conductor} \qquad (33a)$$

The more complete equation is, therefore,

$$p = 241(f + 25)\sqrt{\frac{r + \frac{6}{s} + 0.04}{s}}\,(e - e_d)^2\,10^{-5}\ \text{kw./km. of conductor}^{1} \quad (34a)$$

In Fig. 127 the points are measured corona loss values, while the curve is calculated from equation (34*a*) ($r = 0.032$, spacings 46 and 275 cm.). Data are given in the appendix.

It is interesting to note from (33*a*) that as r becomes very small the c^2 term, at a given spacing, approaches a constant value—somewhat as if the corona diameter acts as the conductor diameter.

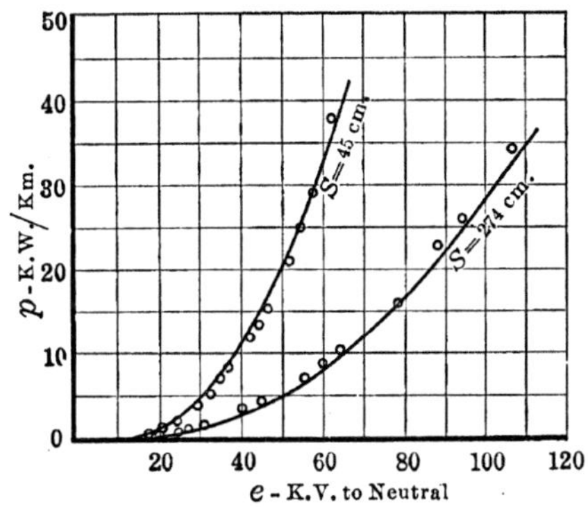

FIG. 127.—Comparison of calculated and measured corona loss for small conductors at small and large spacings. (Points measured. Curves calculated from equation (34*a*) Conductor radius, 0.032 cm. Spacings, 46 cm. and 275 cm. $\delta = 1.01$.)

The Disruptive Critical Voltage.—The point of greatest stress around a cylindrical conductor is at its surface. When s/r is large the gradient at the surface of the conductor may be expressed

$$g = \frac{de}{dx} = \frac{e}{r \log_e s/r} \quad (12c)$$

where e = the voltage to neutral.

s = distance between conductor centers.

r = the radius.

When r and s are in centimeters and e is in kilovolts, g is expressed in kilovolts per centimeter. If e_o, which has been called the *disruptive critical voltage*, is taken for e,

$$g = \frac{e_o}{r \log_e s/r}$$

g_o then is the stress at the conductor surface corresponding to e_o, and will be called the *disruptive gradient*, to distinguish it from the visual gradient g_v. Values of g_o for wires and cables taken under a variety of conditions on the outdoor line are given in Tables LVIII and LIX. These values are corrected to standard temperature and pressure by dividing by δ.

[1] F. W. Peek, Jr., High Voltage Engineering Journal, Franklin Institute, Dec., 1913.

TABLE LVIII.—DISRUPTIVE CRITICAL VOLTAGE GRADIENT FOR WIRES (Values Corrected to 76 cm. Barometer and 25 deg. C., Outdoor Line)

Test No.	Spacing, cm.	Radius, cm.	g. kv./cm. max.	Per cent. variation from mean	Per cent. variation max. to min.
91	152.0	0.084	31.3	....	....
94	229.0	....	31.6	....	....
92	410.0	....	29.1	....	....
95	550.0	....	36.5	....	....
			Avg. = 30.9	5.8	7.9
134	122.0	0.164	28.8	....	....
135	183.0	....	27.1	....	....
136	244.0	....	29.0	....	....
137	366.0	....	25.7	....	....
138	488.0	....	25.3	....	....
			Avg. = 27.2	7.0	12.7
125	91.4	0.259	28.7	....	....
126	183.0	....	26.5	....	....
127	275.0	....	26.0	....	....
128	397.0	....	26.2	....	....
			Avg. = 26.9	6.7	9.4
122	91.4	0.463	28.7	....	....
123	183.0	....	30.4	....	....
120	214.0	....	30.5	....	....
124	275.0	....	31.0	....	....
			Avg. = 30.1	4.8	7.6

Total Avg. 29.0

If the values of e_o for standard line A (1.8 cm. seven-strand cable) are examined it is found that the average value of g_o is 25.8 kv. per centimeter maximum. For cables between 0.583 cm. and 1.18 cm. in diameter and various spacings, the average value of g_o is 25.6 kv. per centimeter maximum, or, in other words, g_o is constant for *commercial* sizes of cables, at practical spacings, and is 25.6 kv. per centimeter maximum. In determining the value g_o for seven-strand cables r was taken for convenience as the outside radius. Hence the above g_o is not the actual g_o as obtained for wires, but is an apparent g_o. The actual g_o may be obtained by taking some mean radius r_2 between the outside radius r and the radius to the point of contact of the outside strands r_1. r_2 approaches r in value as the number of strands is increased.

TABLE LIX.—DISRUPTIVE CRITICAL VOLTAGE GRADIENT FOR CABLES (Values Corrected to 76 cm. Bar. and 25 deg. C., Outdoor Line)

Test No.	Spacing, cm.	Radius, cm.	g. kv./cm. max.	Per cent. variation from mean	Per cent. variation max. to min.
73	91.4	0.292	26.5	....	....
80	152.0	....	24.0	....	....
79	244.0	....	23.9	....	....
77	310.0	....	23.9	....	....
82	432.0	....	25.0	....	....
			Avg. = 24.7	7.3	9.8
100	91.4	0.476	25.5	....	....
115	91.4	....	26.2	....	....
116	183.0	....	26.0	....	....
117	275.0	....	26.4	....	....
118	366.0	....	28.1	....	....
			Avg. = 26.4	6.4	9.3
		Line A	— 1. 2. 3. 4		
18	310.0	0.590	25.5	....	....
36	310.0	....	26.0	....	....
37	310.0	....	25.3	....	....
84	310.0	....	25.8	....	....
101	310.0	....	26.5	....	....
103	310.0	....	26.1	....	....
104	310.0	....	25.7	....	....
105	310.0	....	26.0	....	....
109	310.0	....	25.8	....	....
119	310.0	....	25.1	....	....
119	310.0	....	26.0	....	....
			Avg. = 25.8	2.7	5.3
		Total	Avg. = 25.7	5.5	8.1

The values of g_o for wires varying in diameter from 0.168 cm. to 0.928 cm. and for spacings from 90 to 600 cm. are constant within the limits of experimental error. This more than covers the commercial range. The mean max. value is:

g_o = 29 kv. per cm. at 25 deg. C., and 76 cm. barometric pressure.

Considerable variation should be expected in g_o values obtained on a long outdoor line, due to

1. Necessarily imperfect conductors, kinks, etc., in an outdoor line of this length.

2. Progressive change in the value of successive points on a given curve due to slight changes of wave shape, etc., as the voltages are increased, and the apparent shift of e_o.

The close agreement of g_o for wires is for the above reasons remarkable.

Discrepancies due to progressive change are not to be expected for standard line A to any great extent as the conductor spacing was always the same, and test conditions were kept as nearly constant as possible.

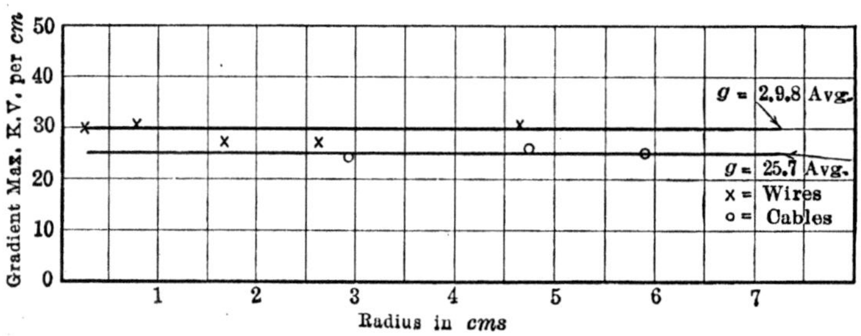

FIG. 128.—0.90 for wires and cables. (Data from Tables LVIII and LIX.)

The above data shows that g_o may be considered constant for diameters of conductors and spacings within the practical range. This immediately suggests that g_o is the actual rupturing gradient of air or identical with g_o of the expression for visual corona. This was further investigated over a greater range on the indoor line.

$$\frac{g_o \text{ cables}}{g_o \text{ wires}} = m_o$$

where m_o is a fraction which approaches unity as the irregularity of the surface is reduced, or number of strands increased. Then

$$e_o = \delta m_o g_o r \log_e s/r \qquad (35)$$

(See curve, Fig. 128.)

The loss equation may now be written

$$p = \frac{241}{\delta}(f + 25)\sqrt{r/s}(e - e_o)^2 \, 10^{-5} \text{ kw./km.} \qquad (34)$$

where p expresses the loss above the visual critical voltage e_v, and e_o is given in (35).

Investigations on the indoor line over a range covering small conductors show that the disruptive gradient may be considered for all practical purposes, constant when the conductor radius is greater than 0.20 cm. As the size of the conductor is decreased the disruptive gradient increases very rapidly. If the disruptive gradient for any size of conductor is called g_d we have found that this law may be expressed:

$$g_d = g_o\delta\left(1 + \frac{0.30}{\sqrt{\delta r}}\frac{1}{(1 + 230\,r^2)}\right) \quad (36)$$

$$e_d = g_d\, r \log_\epsilon \frac{s}{r} \quad (35a)$$

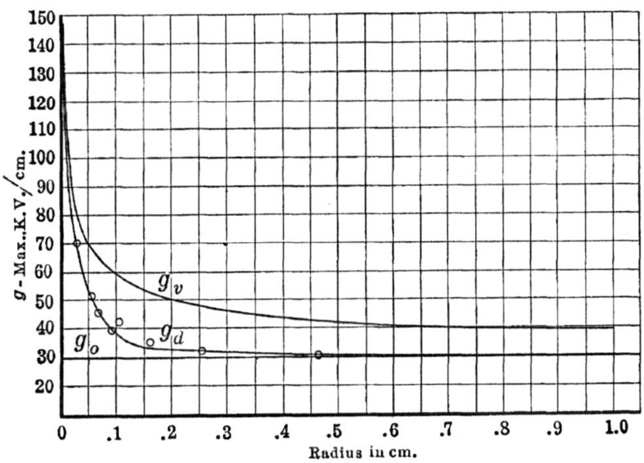

FIG. 129.—Comparison of g_v, g_d, and g_o. (g_d values from Table LX.)

Thus, when r is large (above 0.2 cm.) practically $g_d = g_o$, and where $r = o$, $g_d = g_v$.

This *apparent* increase in disruptive gradient for small conductors gives the impression that the residual from the very highly ionized air resulting at each cycle in effect increases the size of the conductor. Thus, the g_d formula is the same as the g_v formula with the energy distance term apparently modified or lessened by the residual ionization at each cycle. g_o results for large conductors. Measured values of g_d for different conductors on the indoor line are given in Table LX. Calculated and measured values are plotted in Fig. 129.

The corona loss equation over a greater range of conductor diameter and spacing may thus be written

$$p = 241(f + 25)\sqrt{\frac{r + \frac{6}{s} + 0.04}{s}}\,(e - e_d)^2\,10^{-5}\,\text{kw./km.} \quad (34a)$$

where e_d is obtained from equation (35*a*).

TABLE LX

Test No.	r cm.	g_d maximum measured	g_d maximum calculated
12 B	0.032	67.5	70.7
13 B	0.032	70.3	70.7
14 B	0.032	71.0	70.7
18 B	0.032	71.0	70.7
15 B	0.032	71.0	70.7
16 B	0.032	71.0	70.7
19 B	0.032	69.6	70.7
17 B	0.032	71.2	70.7
20 B	0.032	69.5	70.7
21 B	0.032	69.3	70.7
		Avg. =70.0	
27 B	0.057	51.7	51.3
26 B	0.057	51.3	51.3
25 B	0.057	51.7	51.3
24 B	0.057	51.3	51.3
		Avg. =51.5	
30 B	0.071	45.4	45.4
33 B	0.0914	39.4	39.4
32 B	0.0914	39.4	39.4
42 B	0.105	42.2	37.6
41 B	0.105	42.2	37.6
37 B	0.105	42.5	37.6
		Avg. =42.3	
45 B	0.164	36.5	33.0
44 B	0.164	33.6	33.0
		Avg. =35.1	
49 B	0.256	30.8	31.1
48 B	0.256	32.1	31.1
47 B	0.256	33.2	31.1
		Avg. =32.0	
56 B	0.464	31.8	30.2
55 B	0.464	31.8	30.2
54 B	0.464	31.5	30.2
		Avg. =31.7	

It may be interesting to note that equation (34) may be written in terms of the gradient, thus

$$p = A^1\sqrt{r/s}\,(f + 25)\left(\log \frac{s}{r}\right)^2 (r^2)\,(g - g_o)^2 \times 10^{-5}$$

Fig. 130 shows measured curves plotted between g and p for a given wire at three different spacings. These curves all intersect the axis at g_o max. $= 30$, at $\delta = 1.0$.

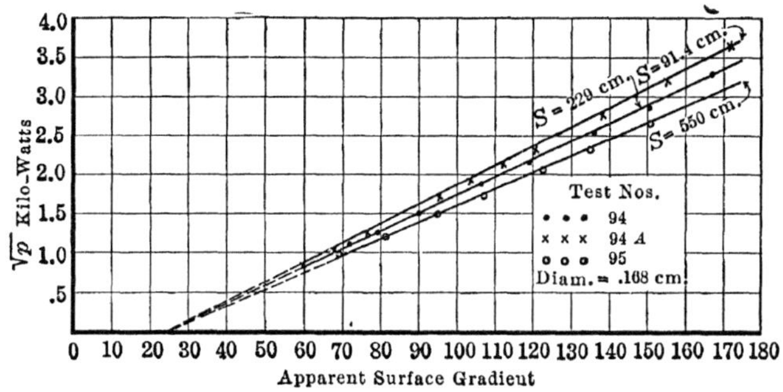

FIG. 130.—Relation between power loss and apparent surface gradient.

Losses Near the Disruptive Critical Voltage—e_o.—If the conductors could be made perfect no appreciable loss would occur below the visual critical voltage. However, near the starting point of corona two effects occur, which cause a deviation of the loss from the quadratic law, equation (34), and which affect the loss in opposite directions:

(*a*) With perfect conductors loss of power does not begin at the voltage e_o, at which the disruptive gradient is reached at the conductor surface, but only after the disruptive strength of air has been exceeded over a finite and appreciable distance a from the conductor, that is, at a higher voltage e_v. Since the convergency of the lines of dielectric force is great at the surface of small conductors, with such conductors, a considerable increase of the voltage is required to extend the disruptive gradient to some distance from the conductor, and e_v is considerably higher than e_o. With such conductors there would be no loss until e_v were reached. The loss would then suddenly take nearly the definite value calculated for this applied voltage from equation (34). Even with *polished* conductors, in practice the decrease of loss below that given by equation 34 is appreciable with small conductors within the range between e_o and e_v, as seen in Fig. 131.

With large conductors, however, the less convergency of the lines of dielectric force at the conductor surface requires a less voltage increase beyond e_o to extend the disruptive gradient to some distance from the conductor; e_o and e_v are therefore closer together, and this decrease of the loss below the theoretical value given by equation (34) is not appreciable.

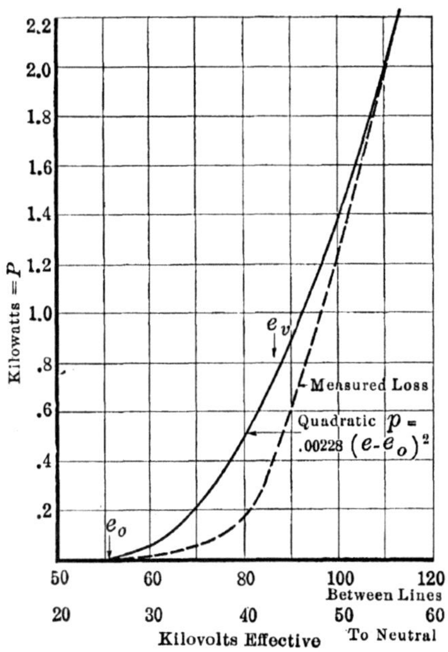

FIG. 131.—Corona loss for small conductors near the critical voltage.

No. 8 copper. Diameter, .328 cm. Length, 29,050 cm. Spacing, 122 cm. Temp. 1.5. Bar., 76.6. Line B.

(*b*) As the conductor surface can never be perfect, some loss of power occurs at and below the disruptive critical voltage at isolated points of the conductor, where irregularities of the surface, scratches, spots of mud or dirt, etc., give a higher potential gradient than that corresponding to the curvature of the conductor surface. With small conductors, this loss is rarely appreciable, since the curvature of the conductor surface is of the same magnitude as that of its irregularities. It becomes appreciable, however, for larger conductors, as seen in Fig. 132. This excess of the loss beyond that given by the quadratic law equation essentially depends on the conductor surface, and is the larger the rougher or dirtier the surface is.

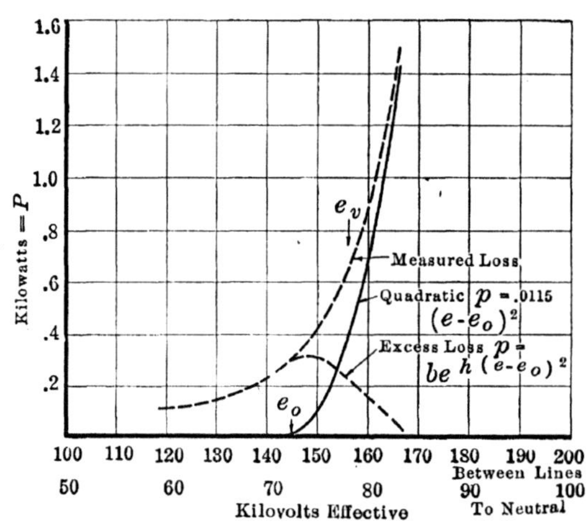

FIG. 132.—Corona loss for large conductors near the critical voltage. (Data same as Fig. 115.)

It is a maximum at the disruptive critical

voltage e_o, and decreases above and below e_v, and is with fair accuracy represented by the probability curve:

$$p = q\epsilon^{-h(e_o - e)^2} \qquad (37)$$

where q is a coefficient depending on the number of spots, and h is a coefficient depending upon the size of spots.

Snow, sleet and rain losses seem to be of the same nature, but frequently of far greater magnitude.

Equation (37) is probably of no practical importance as the loss is small, and q and h naturally cover a wide range of values, depending upon the condition of the conductor surface. Experimental values near e_o are taken from Table XLIV and tabulated in Table LXI, together with values calculated by the quadratic law. Corresponding experimental and calculated values are subtracted and also tabulated.

TABLE LXI.—LINE A 1-2-3-4 FROM TABLE XLIV, TEST 36
$c^2 = 0.0115$ $e_o = 72.1$

Kilovolts between conductors e'	Kw. exp. p_o	Kw. $p = 0.0115(e - e_o)^2$	Excess loss $p_l = (p_o - p)$	$\log_e p_l$	$(e_o - e)^2$
120.4	0.11	....	0.11	2.40	146.5
130.2	0.15	....	0.15	2.71	49.0
139.2	0.22	....	0.22	3.09	6.2
150.0	0.40	0.10	0.30	3.40	8.5
159.0	0.79	0.63	0.16	2.77	54.7
165.8	1.42	1.34	0.08	2.10	11.7

Fig. 132 is plotted to a large scale to show the excess loss near e_o. As this is for large conductors, e_o and e_v are near together, and the effect of (b) predominates. In order to see if equation (37) holds, write

$$\log_e p_l = \log q - h(e_o - e)^2$$

Then the curve between $\log p_l$ and $(e_o - e)^2$ should be a straight line. This is shown in Fig. 133.

Values of q and h are of the following order for Line A:

Test No.	q per cm. total conductor	h
30	3.19×10^{-6}	−0.0220
105	2.47×10^{-6}	−0.0208
103	2.74×10^{-6}	−0.0304

Fig. 131 is plotted from values for a small smooth conductor. Here e_o and e_v are far apart, and as the curvature of the conductor surface is of the same magnitude as its irregularities they do not greatly influence the loss. The (*a*) effect here predominates—that is, the loss near e_c is lower than that shown by the quadratic law.[1]

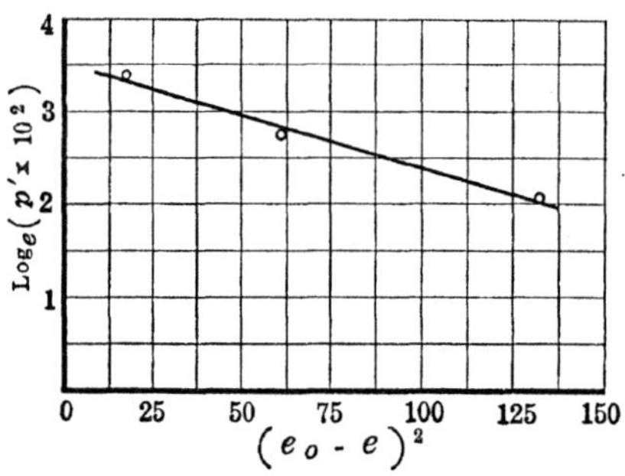

FIG. 133.—Determination of equation (37).

The loss between e_o and e_v is unstable as it depends upon surface conditions. In *all* cases the quadratic law is closely followed above e_v, and on large practical sizes of conductors with sufficient accuracy over the whole range.

Temperature and Barometric Pressure.—Values of the disruptive critical voltage e_o covering a considerable temperature range are tabulated in Table LXII. Correction is made to a

TABLE LXII.—TEMPERATURE AND DISRUPTIVE CRITICAL VOLTAGE e_o
(Standard Line A 1–2–3–4)

Test No.	Temperature		Bar. cm.	e'_o kv. bet. lines	e'_o corr. to 76 cm.	$1/e'_o$	Weather
	Wet	Dry					
18	16.0	18.5	75.5	138.5	140.0	0.00715	Bright sun
15	20.0	22.0	75.2	140.0	142.0	0.00705	Cloudy sun
37	10.0	13.0	75.7	141.0	142.0	0.00705	Bright sun
36	10.0	12.0	75.0	144.2	146.5	0.00683	Cloudy
109	−3.0	−2.0	73.9	148.8	150.8	0.00663	Hazy sun
84	1.0	3.0	75.2	149.0	151.0	0.00662	Cloudy
101	−1.0	−1.0	74.7	153.8	156.8	0.00638	Cloudy
103	−4.9	−4.5	75.7	155.3	156.7	0.00638	Sun
104	−9.5	−9.5	76.2	157.0	157.0	0.00637	Sun
119	−6.5	−6.0	76.5	156.6	156.1	0.00642	Sun
105	−13.0	−13.0	76.2	161.0	161.0	0.00622	Sun

barometric pressure of 76 cm. on the assumption that e'_o varies directly with the pressure. In Fig. 134 $1/e'_o$ is plotted with temperature. The straight line through these points cuts the temperature axis at −273 deg. C. or absolute zero. Temperature was always measured in the shade. The points that do not fall

[1] This condition is likely to obtain to a greater extent at high altitudes as e_o and e_v are farther apart.

well on the curve are the summer sunny day points. This is what would be expected, as the conductors were at a higher temperature than the temperature read.

Fig. 134 shows that the disruptive critical voltage or the disruptive gradient varies inversely as the air density. The data range, however, is not great.

The density of air at 25 deg. C. and 76 cm. barometric pressure is taken as the standard.

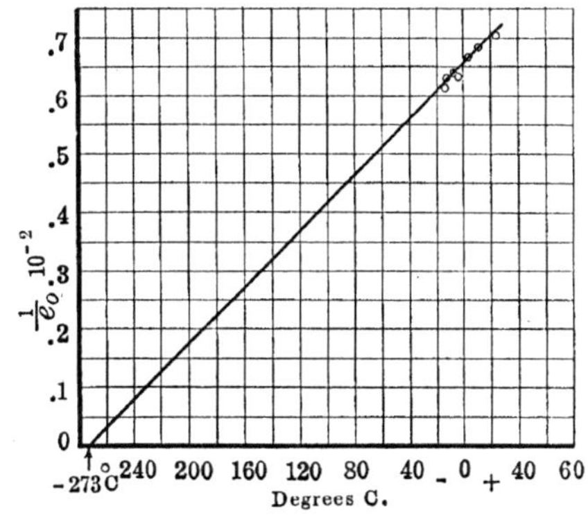

FIG. 134.—Effect of temperature on e_o. (Data from Table LXII.)

Humidity, Smoke, Wind. *Humidity.*—Line A was kept as a standard throughout the tests. A careful study of the disruptive critical voltage and c^2 shows no appreciable effect of either humidity or "vapor products."[1] (See Table LXIII.) Visual tests, made on two short parallel wires indoors over a great humidity range, also bear this out.

TABLE LXIII.—STANDARD LINE A.[2] CONDUCTORS 1-2-3-4

Test No.	Temperature		Relative humidity	Vapor product	g_o reduced to 25 deg. C., 76 cm. bar.
	Wet	Dry			
15	20.0	22.0	0.84	0.55	18.8
18	16.0	18.5	0.75	0.35	18.5
36	10.0	12.0	0.78	0.25	18.8
37	10.0	13.0	0.67	0.21	18.3
84	1.0	3.0	0.69	0.11	18.7
101	−1.0	−1.0	1.00	0.17	19.2
103	−4.9	−4.9	1.00	0.13	18.9
104	−9.5	−9.5	1.00	0.08	18.6
105	−13.0	−13.0	1.00	0.06	18.8
109	−2.0	−2.0	1.00	0.12	18.7

Theoretically there should be an appreciable effect of humidity

[1] Mershon, High Voltage Measurements at Niagara, A.I.E.E., June 30, 1908.

[2] Line A was kept at constant test conditions for use as a standard in the study of varying atmospheric conditions, etc.

on e_o, since even if the water vapor, which may be considered as a gas dissolved in air, has a different disruptive gradient than air, the percentage of the gas in the mixture should be too small to cause any appreciable change. It has been suggested that "vapor products" is a measure of ionization and, in that way, the critical voltage varies with vapor products. This does not seem likely, because with all ordinary atmospheric air the percentage of ionization is so small that it would not be expected to produce any effect. To test this, the visual critical point was determined on two parallel wires. The room was then closed and the wires were run at a point very much above this critical point for about an hour, or until a very intense odor of ozone filled the room. The voltage was then removed and the surface of the wires cleaned in order to remove oxidization. The critical point was then redetermined and found to be the same, although the amount of ionized air was many times that which could be expected in free atmospheric air. Of course if the percentage of ionization were great enough, as for instance in an ozone machine, a change in the disruptive strength would then be expected.

It has been claimed that ultraviolet light reduces the sparking point. This is not borne out in tests, where any quantity of continuously applied power is involved. Though ultraviolet light, ionized air, and various radiations cause the small energy in condensers to discharge when applied over comparatively great time; in case of large energy discharge in a very short time, as spark discharge, and corona due to continuously applied voltages, no appreciable effect should be expected, or can be observed, since the discharge which takes place by ionized air is not of the same order of magnitude as that produced by the spark discharge. The time required for a given voltage to produce ionic saturation or discharge, should, however, be less with high initial ionization than with low initial ionization. Hence for impulse voltages of steep wave front and short duration, the spark voltage may be affected by initial ionization. This is further discussed in Chapters IV and VIII (page 198).

It is quite probable that humidity has some slight effect on the loss after corona has once formed due to the change of the gas into vapor and the agglomeration of the water particles by the ions. This is noticed in the spark-over between needle points at very high voltage. In this case there is a very heavy brush discharge before spark-over. The sparking voltage increases with increasing humidity, due to the fog formed, when there is not a mixture of two gases, air and water vapor, as in the case where humidity is concerned, but actual water particles are in the air. Steinmetz has observed that fog actually raises the striking distance between needle points. The effect of humidity on the spark discharge observed by the author is discussed in Chapter IV. Greater loss should be expected in corona measurements during fog due to charge and discharge of the water particles. This causes loss at lower voltages and has the effect of decreasing the critical point.

While humidity has no effect on the starting point of corona, an effect might be expected on the loss after the discharge had already started. The reason that this is inappreciable is because the corona discharge from wires covers very little space.

Smoke.—It was difficult to get measurements to show the effect of smoke, as the prevailing winds were over the fields toward the city. At one time, however, during a change in the wind thick smoke was blown over the line from smoke stacks of a factory, and the loss was increased. This, however, will probably not be a serious consideration in practice.

Wind.—Losses measured during very heavy winds show no variation from losses measured during calm weather. That the losses do increase with increasing air velocity has been shown in laboratory apparatus. There is, however, no appreciable effect due to the comparatively low air velocity in practice.

Moisture, Frost, Fog, Sleet, Rain and Snow.—During some of the first tests it was noted that the losses were sometimes greater on the "going-up curve" than on the "coming-down curve," especially in the early mornings after heavy dew. The losses became less after the line had been operated for a while at high voltage. Fig. 135 shows this well for a conductor with a coating of frost. This excess loss was thought at first to be due to leakage through moisture on the insulators. Insulators were put up without line wires, but measurements showed a very small insulation loss even during storms. It was then concluded to be due to moisture on the conductors themselves. Visual tests made on short lengths of wet and dry cables showed this in a very striking manner. Two parallel dry cables were brought up to the critical point. Water was then thrown on the cables. What had been a glow on the surface of the dry cables now became, at the wet spots, a discharge extending as much as 5 to 8 cm.

from the cable surface. This discharge reminded one of an illuminated atomizer. Illustrations, Figs. 67 and 68 show this, but a greater part of the effect is lost in reproduction. The wires became quite dry and down to normal discharge after running at high voltage for a very few minutes.

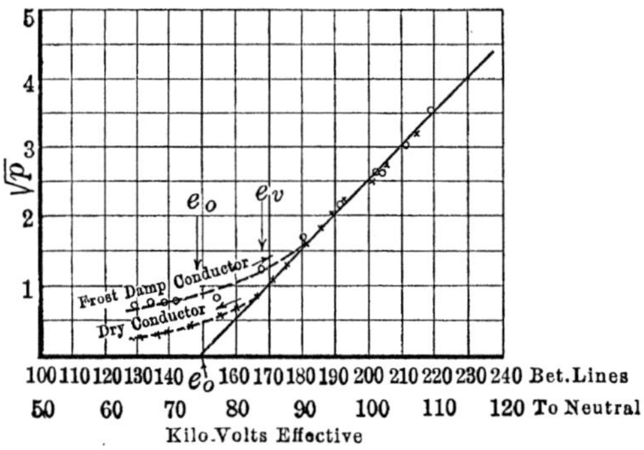

FIG. 135.—Corona loss with frosty and dry conductors.

(Conductor length, 109,500 cm. Spacing, 310 cm. Diameter, 1.18 cm. 3/0,7-strand cable. Line A. Temp., −2° C. Bar., 74 cm.)

The curves, Fig. 136, taken during the fog also show the combined effect of condensed moisture on the cables, and free water particles in the air. The moisture particles on

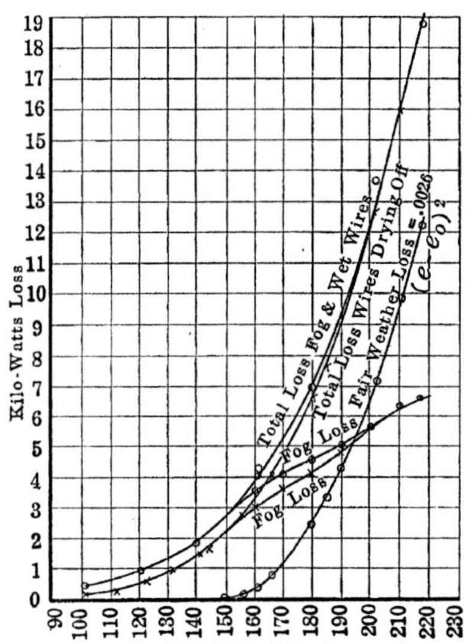

FIG. 136.—Corona loss during fog; conductors wet.

(Conductor length, 109,500 cm. Diameter, 1.18 cm. Spacing, 310 cm. 3/0,7-strand cable. Line A. Temp., 2. Bar., 75.5.)

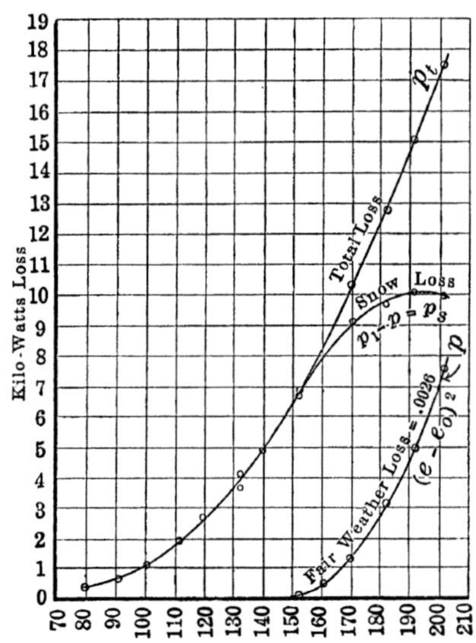

FIG. 137.—Corona loss during snow storm.

(Conductor length, 109,500 cm. Diameter, 1.18 cm. Spacing, 310 cm. 3/0,7-strand cable. Line A. Temp., 0. Bar., 74.2.)

the conductor become charges and are repelled. The particles in the air also become charged and discharged thus increasing the loss very greatly above that for dry conductors.

The losses during snow and rain storms are much greater than fair weather losses at the same temperature and barometric

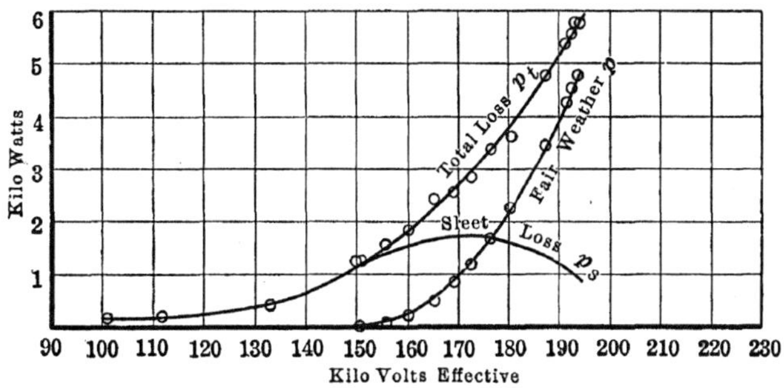

FIG. 138.—Corona loss during sleet storm.
(Conductor length, 109,500 cm. Diameter, 1.18 cm. Spacing, 310 cm. 3/0,7-strand cable. Line A. Temp., – 1.0. Bar., 75.4.)

pressure. In Fig. 137 the actual measured loss is plotted, and also a corresponding calculated fair weather loss. The difference between the two curves shows the excess loss due to snow. The effect of snow is greater than that of any other storm condition. This is because the particles are larger and a greater number strike the line, or come near the line.

The sleet curves are of special interest. Sleet had already started to form on the conductors, and was still falling when the tests were started. Fig. 138 shows the loss curves. After the curves were taken the line was kept at 200,000 volts for over an hour with no apparent diminution of sleet. This seems to show that sleet will form on high-voltage transmission lines.

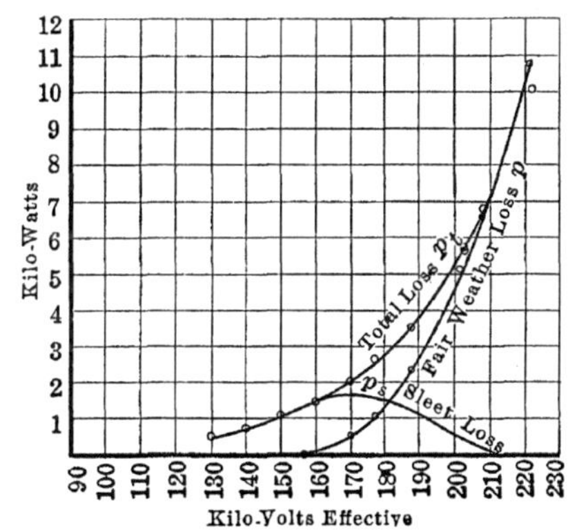

FIG. 139.—Corona loss with sleet on the wires.
(Conductor length, 109,500 cm. Diameter, 1.18 cm. Spacing, 310 cm. 3/0,7-strand cable. Line A. Temp. 10.0. Bar, 76.)

The day after these tests were made was bright and clear and the conductors were still coated with sleet. A set of readings was taken, and it is interesting to note that the excess loss here is as great as when sleet was falling. (See Fig. 139.)

The excess loss for sleet, rain or snow storms (over the fair weather loss) seems with increasing voltage to approach a maximum and then to decrease again (the latter at a value very far above the disruptive critical voltage), and the curves of loss seem to have the general shape of the probability curve, as is to be expected theoretically.

The above readings show the importance of taking weather conditions into account in the design of high-voltage transmission lines.

Very High Frequency.—Corona losses at very high frequency are difficult to measure. The curve in Fig. 140 is interesting. The drawn curve is calculated from formula 34(a); the points are measured values. Power was supplied from an Alexanderson 100,000-cycle alternator. The power was measured by adjusting reactance and capacity until unity power factor was obtained. The watts input was then the product of volts and amperes as measured by hot wire meters.

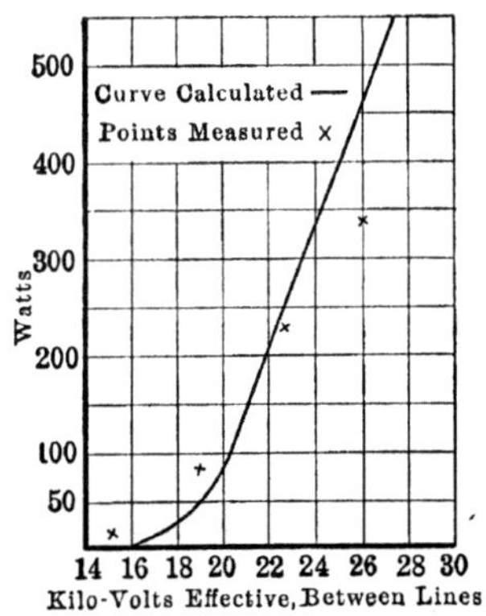

FIG. 140.—Corona loss at high frequency.

(Two parallel wires. Spacing, 67 cm. Radius of wire, 0.127 cm. f = 100,000 ∼. Total length, 200 cm.) Tests by Alexanderson.

This good check seems to show that the formula applies over a large range, but complete conclusions cannot be drawn from this small amount of data.

CHAPTER VI

CORONA AND SPARK-OVER IN OIL AND LIQUID INSULATIONS

The most common liquid insulation is transformer oil obtained by fractional distillation of petroleum. This oil has various characteristics, as flashing point, freezing point, viscosity, etc., depending upon the specific use to which it is to be put. The average characteristics of transformer oil are as follows:[1]

	Medium	Light
Flashing temperature............	180°–190° C.	130°–140° C.
Burning temperature............	205°–215° C.	140°–150° C.
Freezing point..................	−10°– −15°	−15°– −20°
Specific gravity at 13.5 deg. C....	0.865–0.870	0.845–0.850
Viscosity at 40 deg. C. (Saybolt test)	100–110 sec.	40–50 sec.
Acid, alkali, sulphur, moisture....	None	None

Various other oils are insulators, as gasoline and cylinder oil; animal oils, as fish oil; vegetable oils, as linseed oil, nut oil, china wood oil, etc. All of these when pure have the same order of dielectric strength.

The so-called compounds made by dissolving solid gums in oil to increase viscosity are generally unreliable, unless used dry as varnish. Under the dielectric field the dielectrics of different permittivities tend to separate, and there is considerable loss. As in air there is very little loss in pure oil until local rupture, that is, brush discharge or corona, occurs.

The dielectric strengths of oils are usually compared by noting the spark-over voltage between two parallel brass discs 1.25 cm. in diameter, and 0.5 cm. separation. The spark-over voltage for good oils in the above gap should be between 50 and 70 kv. maximum, and hence, if the voltage wave is a sine, from 35 to 50 kv. effective.[2]

[1] Tobey, Dielectric Strength of Oil, A.I.E.E., June, 1910.

[2] The breakdown voltage of oil like that of air depends upon the maximum point of the wave.

Different Electrodes.—In Fig. 141 are plotted spark-over curves for different electrodes in good transformer oil. The characteristics are very much the same as for air excepting the apparent strength is very much higher.

Effect of Moisture.—The slightest trace of moisture in oil greatly reduces its dielectric strength. The effect of moisture is shown in Fig. 142(*a*) for the standard disc gap. Water is held in suspension in oil in minute drops. When voltage is applied

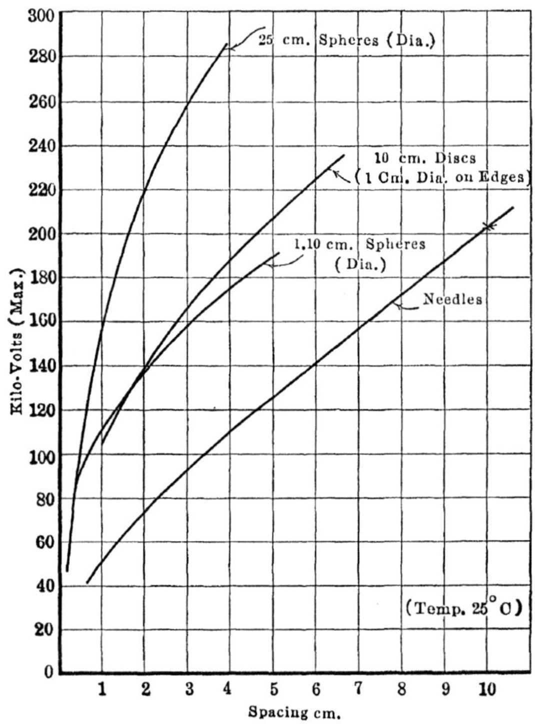

FIG. 141.—Spark-over of various shaped electrodes in oil.

these drops are attracted by the dielectric field. Thus they are attracted to the denser portions of the field and may form larger drops by collision. When attracted to, and after touching a metal part, and thus having the same potential, they are immediately repelled. If the field is uniform the drops form in conducting chains along the lines of force. It can be seen that the effect of moisture should vary greatly with the shape of the electrode, and with some shapes the moisture may even be removed from the space between the electrodes by the action of the field, in which case its presence would not be detected by low-voltage breakdowns. In transformers moisture will generally be attracted to points under greatest stress. The most effective way of removing moisture is by filtration through blotting paper. Dirt in oil may have an effect very similar to moisture and the small conducting particles be made to bridge between electrodes by the dielectric field.

Temperature.—Temperature over the operating range has very little influence on the strength of oil. The strength increases at the freezing point. The curve is shown in Fig. 142(*b*). The insulation resistance is also shown. The increase in strength

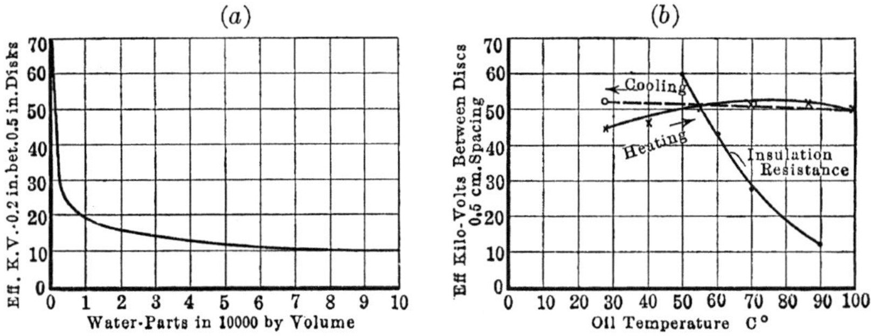

FIG. 142.—(*a*) Effect of water in oil. (*b*) Effect of temperature on dielectric strength and insulation resistance.

with temperature is only apparent and due partly to the decreasing insulation resistance which allows more current to flow through the oil, which tends to even up the stress, but mostly to the drying out of moisture particles by the high temperature. The increase at freezing should be expected, as an actual improvement in dielectric properties results. For a perfectly dry oil the strength actually decreases with increasing temperature or density, as in the case of air.

The specific resistance of transil oil is approximately:

Temp., C	Ohms, cm. cube.
30	250×10^{10}
60	100×10^{10}
100	15×10^{10}

The permittivity is approximately 2.6 times its specific gravity and thus decreases with increasing temperature.

Spark-over and Corona in Oil.—A phenomenon similar to corona in gases also takes place in liquid insulations, as oil, due to a tearing apart of the molecules. Corona in oil is not as steady or definite as in air. It appears to start quite suddenly and extend much farther out from the electrode than corona in air. It is much more difficult to detect the starting point, and unless the conductors are very small or far apart (s/r large) corona does not appear before spark-over. For instance, in the Table LXVII when the outer cylinder is 3.81 cm. radius and the inner 0.0127cm. radius ($R/r = 300$), the corona and spark-over voltages are practically the same (see condition for spark-over and corona, Chapter II). The absence of corona, or rather the simultaneous appearance of corona and spark-over, unless the wires are very small or far apart, seems to mean that the mechanism of breakdown in oil is very similar to that in air but the energy distance in oil is much greater. Thus, as the voltage is increased, corona rupture occurs out to the energy distance; this increases r to the condition for spark-over.

$$\frac{R}{(r + \text{energy distance broken down})} = \frac{R}{r_1} < \epsilon$$

and spark follows. Therefore, the spark-over voltages and

TABLE LXIV.—DIELECTRIC STRENGTH OF NO. 6 TRANSIL OIL—SPARK-OVER BETWEEN SPHERES

Spacing, cm.	Radius of spheres, cm.										Needles
	0.159		0.555		1.27		3.12		6.25		0
	Kv. max.	Gradient max. kv./cm.	Kv. max.	Gradient max. kv./cm.	Kv. max.	Gradient max. kv./cm.	Kv. max.	Gradient max. kv./cm.	Kv. max.	Gradient max. kv./cm.	Kv. max.
0.129	44.6	449	48.0	394	47.1	364	.....	.....	.....	.....	.....
0.198	50.5	360	.....	.....	.....	.....	.....	.....	.....	.....	.....
0.264	59.2	365	73.9	333	73.5	310	74.3	288	74.8	289	.....
0.322	65.0	360	.....	.....	.....	.....	.....	.....	.....	.....	.....
0.378	71.1	364	90.0	295	.....	.....	.....	.....	.....	.....	.....
0.508	81.0	368	.....	.....	99.0	222	105.0	220	107.0	214	.....
0.650	.....	.....	.....	.....	.....	.....	.....	.....	.....	.....	41.6
0.766	89.9	353	106.0	225	117.0	187	128.0	182	132.0	180	.....
1.010	97.6	358	112.0	192	.....	.....	146.0	166	159.0	166	.....
1.270	104.0	361	116.0	176	157.0	158	171.0	158	177.0	147	57.0
1.780	111.0	384	131.0	171	165.0	143	203.0	137	214.0	130	.....
2.540	124.0	416	149.0	174	185.0	130	240.0	122	245.0	110	84.0
3.810	145.0	470	172.0	184	206.0	131	266.0	103	280.0	90	108.0
5.080	168.0	542	191.0	192	231.0	133	.....	.....	.....	.....	124.0
7.620	.....	.....	.....	.....	.....	.....	.....	.....	.....	.....	166.0
10.150	.....	.....	.....	.....	.....	.....	.....	.....	.....	.....	203.0

Oil between standard discs 0.5 cm. apart tested 58.5 kv. max. 25 deg. C.

TABLE LXV.—SPARK-OVER BETWEEN PARALLEL PLATES IN NO. 6 TRANSIL OIL

e kv. max.	X spacing cm.	$g_s \cong \frac{e}{x}$ kv./cm. (max.)	Remarks
34.6	0.254	136.3	
52.3	0.508	102.8	
71.0	0.762	93.0	
106.0	1.015	104.8	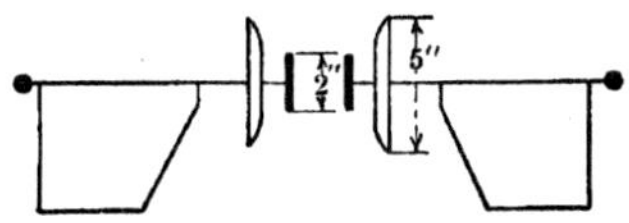
114.0	1.270	89.6	
126.5	1.525	82.8	
			Distance between small and large discs = 0.476 cm.
125.2	1.270	98.2	
137.5	1.525	90.0	
158.0	1.778	85.0	
167.7	2.03	82.2	
187.3	2.29	81.8	
			Same as above, except distance between small and large discs = 1.58 cm.
196.5	2.54	77.2	
214.5	2.79	76.6	
212.7	3.05	69.5	
242.0	3.30	73.2	
257.5	3.56	72.2	
241.0	3.81	63.0	
113.2	1.27	88.8	
155.5	2.54	61.0	
			4-in. flat discs, 0.5-cm. radius edge.
212.0	5.08	41.7	
269.0	7.62	35.0	

TABLE LXVI.—CORONA IN OIL, WIRE AND PLATE
(Distance of wire from plate = 16.5 cm.)

Kv. eff.	Radius wire, cm.	g_v kv./cm. eff.	g_v kv./cm. max.
50	0.025	278	393
60	0.050	185	262
80	0.0635	201	284
100	0.1520	122	173
55	0.00508	615	870

corona voltages up to fairly high ratios of R/r are the same, and may be used in determining the strength of oil.

Table LXVII.—Spark-over Voltages for No. 6 Transil Oil Concentric Cylinders

R cm.	r cm.	Kv. eff.	Kv. max.	g_s max. kv./cm.	$\frac{1}{\sqrt{r}}$	$\frac{R}{r}$	Remarks
3.81	0.032	45.3	64.0	420.0	5.61	120.00	Tests made in long cylinders with belled ends. Oil between standard discs 0.5 cm. apart tested 58 kv. (max.) 25 deg. C.
3.81	0.238	60.0	84.8	127.7	2.05	16.00	
3.81	0.317	60.5	85.5	108.1	1.77	12.00	
3.81	0.635	69.5	98.3	86.3	1.26	6.00	
3.81	0.794	75.0	106.1	85.5	1.12	4.80	
3.81	0.952	73.0	103.2	78.1	1.02	4.00	
3.81	1.111	76.7	108.5	79.4	0.95		
3.81	1.270	76.0	107.5	77.0	0.89	3.00	
3.81	1.587	73.7	104.3	75.1	0.79	2.40	
3.81	1.905	66.3	93.7	70.7	0.72	2.05	
3.81	2.540	45.5	64.3	62.4	0.63	1.57	
6.67	5.560	29.2	41.3	40.6			Tests made on short cylinders where field is somewhat distorted.
6.67	5.080	43.7	61.8	44.8			
6.67	3.970	70.1	99.2	48.1			
6.67	3.240	84.4	119.5	51.0			
6.67	2.230	105.8	149.7	61.2			
6.67	1.955	105.0	148.5	61.9			
6.67	1.615	103.5	146.5	64.0			
6.67	1.270	101.3	143.2	67.5			
6.67	0.953	98.3	139.0	75.0			
6.67	0.635	94.0	133.0	89.2			
6.67	0.477	91.5	129.5	102.8			
6.67	0.318	85.3	120.7	121.0			
6.67	0.159	65.2	92.2	155.5			
6.67	0.079	62.2	88.0	251.4			
11.42	3.930	144.0	203.5	48.3			
11.42	2.280	163.0	230.4	63.0			
11.42	1.910	158.0	223.2	64.5			
11.42	1.270	160.0	226.0	81.5			
11.42	0.880	148.0	209.0	93.0			
11.42	0.520	90.0	127.3	79.5			

The strength of oil for different sizes of wire from Table LXVII is plotted in Fig. 143. The curve is similar to that for corona in air. Fig. 144 shows that a straight line relation holds approximately between $\frac{1}{\sqrt{r}}$ and g_v. Values are not used when $R/r > 3.5$. Thus, as in the case of air:[1]

$$g'_v = g'_o\left(1 + \frac{\propto}{\sqrt{r}}\right)$$

$$g_v = 36\left(1 + \frac{1.2}{\sqrt{r}}\right)\text{kv./cm. max.}$$

$$g_v = 25.5\left(1 + \frac{1.2}{\sqrt{r}}\right)\text{ kv./cm. effective sine wave.}$$

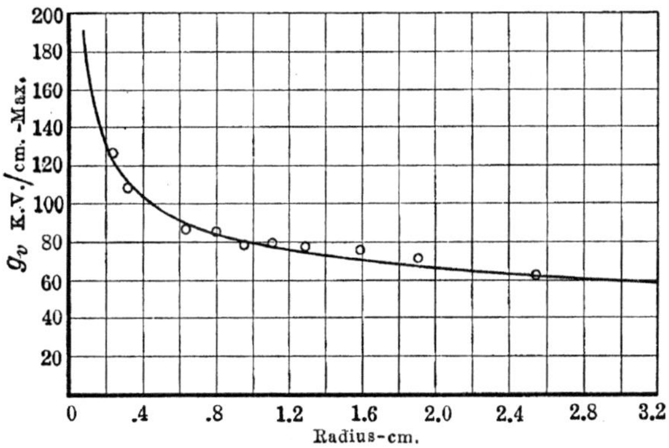

FIG. 143.—Strength of oil.
(Concentric cylinders—g_v at surface of inner cylinder.)

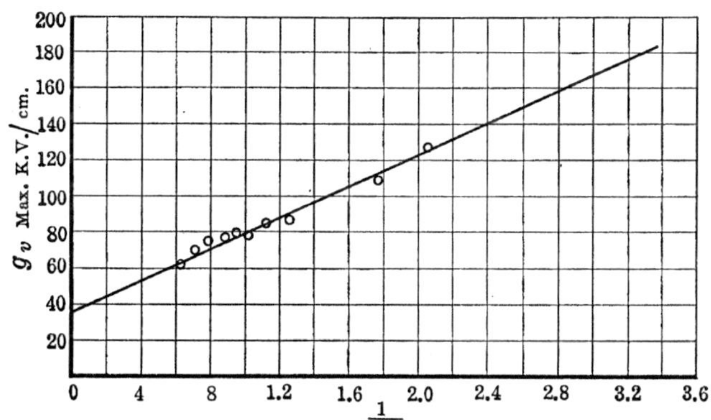

FIG. 144.—Strength of oil.
(Method of reducing values given in Fig. 143 to equation.)

The energy distance is $1.2\sqrt{r}$ or almost four times that of air, indicating that a greater amount of energy is required to rupture oil, or a greater number of collisions are necessary before ionic

[1] F. W. Peek, Jr., High Voltage Engineering, Journal Franklin Institute, Dec., 1913.

saturation is reached. g_o and a vary to a considerable extent in oil. The strength of oil, or the disruptive gradient, or the gradient required to bring the ions up to collision velocity, seems fairly low. Oil should, therefore, have low strength in bulk, but high

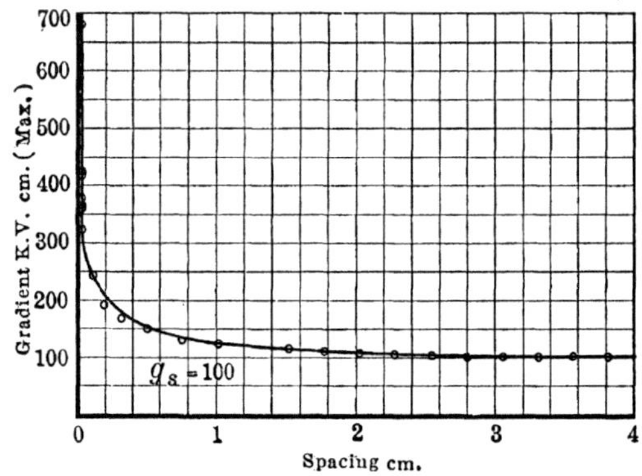

FIG. 145.—Showing increase of strength of oil at small spacings. (Spheres, $R = 3.33$ cm.)

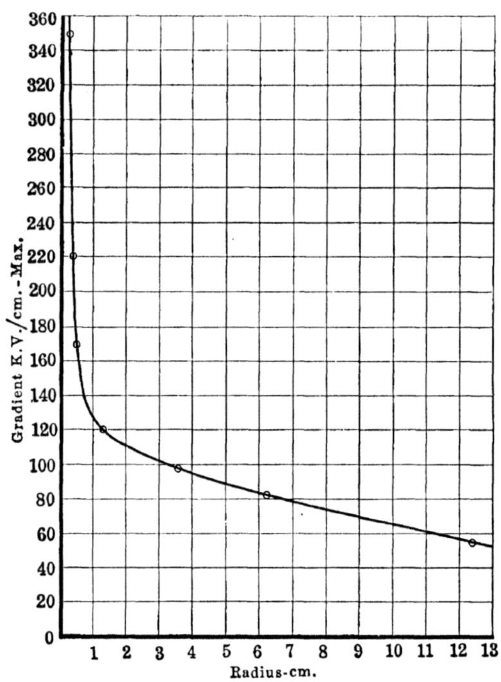

FIG. 146.—Strength of oil between spheres. (Data from constant part of the curve.)

apparent strength when sub-divided or confined to make use of the large energy distance necessary to rupture.

Spark-over voltages and gradients are given for various spheres at various spacings in Table LXIV. The characteristics of the curves between gradient and spacing are the same as those for air (see Fig. 42) as shown in Fig. 145. When the spacing is so small that it interferes with the energy distance the apparent strength of oil increases. At spacings above this the gradient is constant until the separation is so great that corona forms before spark-over.

TABLE LXVIII.—SPHERES IN OIL

Gradient at *constant part* of the curve

Data from Table LXIV

Radius, cm.	Gradient, kv./cm. max.	$\frac{1}{\sqrt{R}}$	Radius, cm.	Gradient, kv./cm. max.	$\frac{1}{\sqrt{R}}$
0.159	348	2.51	1.27	120	0.89
0.237	260	2.05	3.12	98	0.56
0.355	222	1.68	6.25	82	0.40
0.555	169	1.35	12.50	56	0.28

The rupturing gradient at the *constant part* of the curve for various spheres is given in Table LXVIII, and is shown in Fig. 146. This may be written approximately

$$g_s = 28.3\left(1 + \frac{4}{\sqrt{R}}\right) \text{ kv./cm. max.}$$

The energy distance is approximately $2\sqrt{R}$.

The spark-over voltages for concentric cylinders, where $R/r <$ about 5, and for spheres above $2\sqrt{R}$ spacing on the constant part of the curve, may be approximated by substituting g_s in the voltage formula

$$e = g_s r \log_e \frac{R}{r} \text{ cylinders}$$

$$e = g_s \frac{X}{R} f \text{ spheres}$$

In oil, the apparent strength can be greatly improved by limiting the free energy distance by barriers, etc. This can be seen in Tables LXIV and LXV, where e/X is given for parallel planes, and g_v for spheres. For the small spacings, where the free energy distance is limited, the apparent rupturing gradient is very high, just as in the case of air. Strengths as high as 700 kv./cm. have been reached. Insulation barriers give an added effect by preventing moisture particles from lining up.

Care must be taken, however, that the barriers are not so placed as to increase the stress on the oil by the high permittivity of the solid insulation.

Transient Voltages.—Transient voltages or impulse voltages of short duration greatly in excess of the low frequency rupturing voltages may be applied to insulations without rupture. In other words, the rupture of insulation requires not only a suf-

(a)

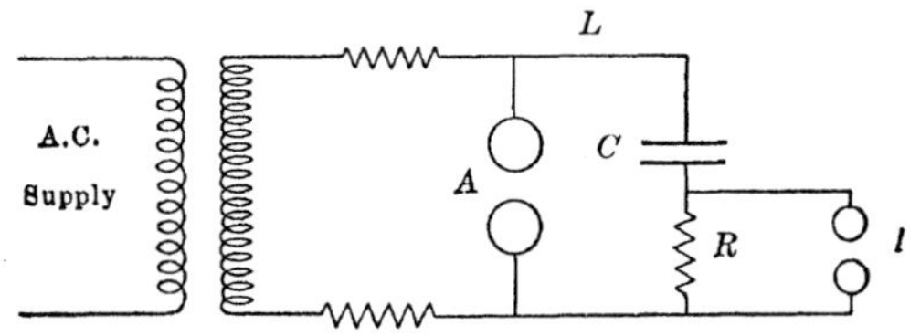

Above circuit used by author. A is sparked-over by transformer voltage. C then discharges through L, arc at A, and R. Impulse appears across R, at I. Arc at A is non-oscillatory and, in effect, a switch.

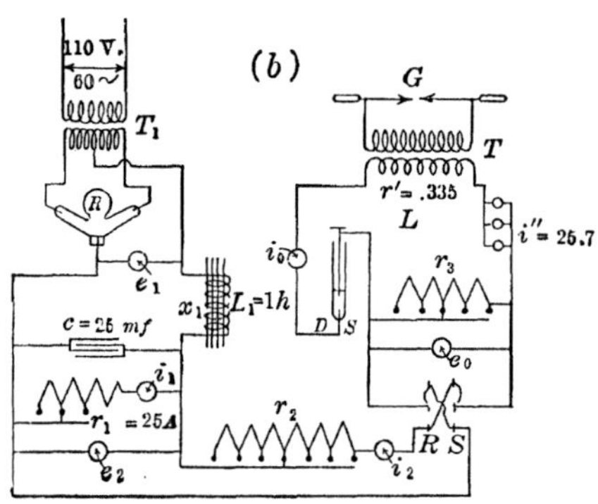

FIG. 147.—"Impulse" circuits.

ficiently high voltage, but also a definite minimum amount of energy. This means also that a definite but very small time elapses between the application of voltage and breakdown. This is sometimes called the "dielectric spark lag" and has already been discussed for air in Chapter IV. The rupturing energy seems to be much greater for oil than for air, as indicated by the large energy distance in the gradient equation above. At low frequency a given definite voltage is required to cause rupture during the comparatively unlimited time of application. This voltage is constant until the application is limited to a definite minimum time, when a higher voltage is required to accomplish the same results in limited time. Such transient voltages of short duration, and impulses of steep wave front, must not be confused with continuously applied high frequency where breakdown will generally take place at lower voltages, due to loss, etc.

In Table LXIX the relative breakdown voltages of gaps in oil, at 60 cycles, and for impulse voltages of steep wave front, are given. An impulse voltage much higher than the 60-cycle voltage is required to break down a given gap. If similar air and oil gaps are set to rupture at the same low frequency voltage, a much higher transient voltage will be required to rupture the oil gap than the air gap, indicating greater time. An air gap may thus protect an oil gap, but not vice versa. See page 108.

Table LXIX.—Comparison of 60-cycle and Impulse Spark-over in Oil and in Air

Oil				Air			
Gap	Spacing, cm.	Kv. 60 cy., max.	Kv. impulse, max.	Gap	Spacing, cm.	Kv. 60 cy., max.	Kv. impulse, max.
Stand. disc....	0.5	56.6	170.0				
2/0 needles...	1.0	50.2	103.3	2/0	5.1	50.2	67.8
	2.0	68.8	157.0	needles	17.5	108.0	198.0
	3.0	89.2	233.3				
	4.0	108.0	321.0				
	0.25	37.2	117.3				
2.54-cm.	0.51	62.8	199.4				
spheres.	0.77	87.5	279.0				
	1.02	111.2	337.0				
6.25-cm.	0.25	47.5	162.5				
spheres.	0.51	68.3	244.5				
	0.70	88.3	267.0				
	1.02	115.8	284.0				

The above voltages are measured by sphere gaps at low frequency and for impulse. The difference between the 60 cycles and impulse voltages increases as the steepness of the impulse increases. Fig. 147 shows impulse circuits. Fig. 147(*b*),

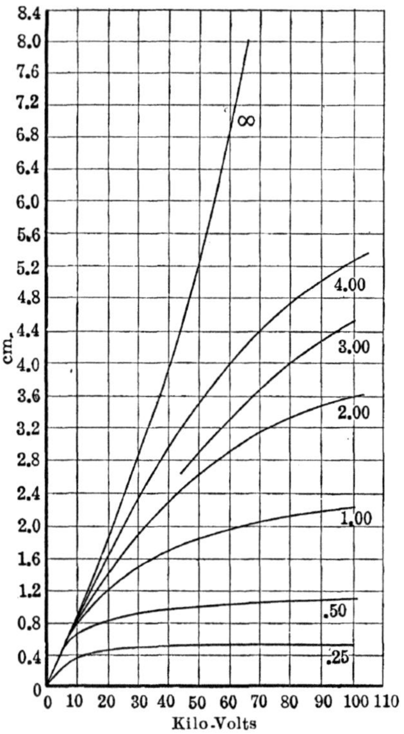

FIG. 148.—Needles in air.

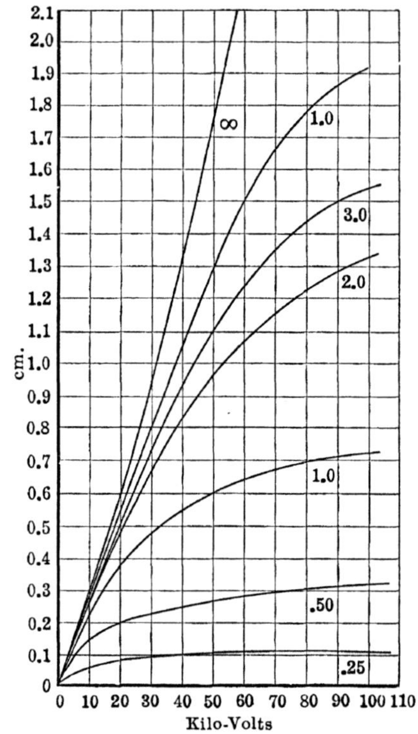

FIG. 149.—Spheres in air.

(Transient voltages. Limited energy.)

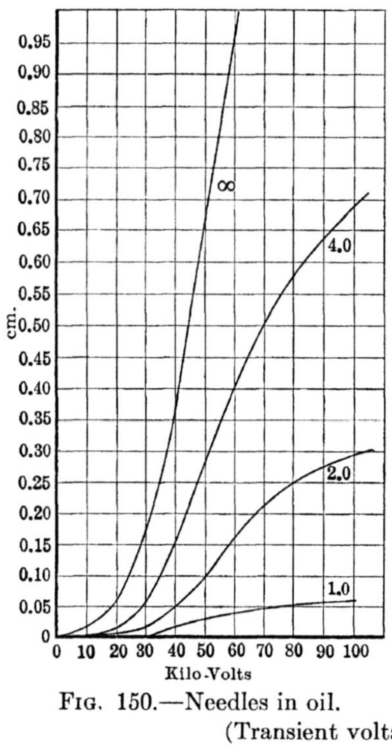

FIG. 150.—Needles in oil.

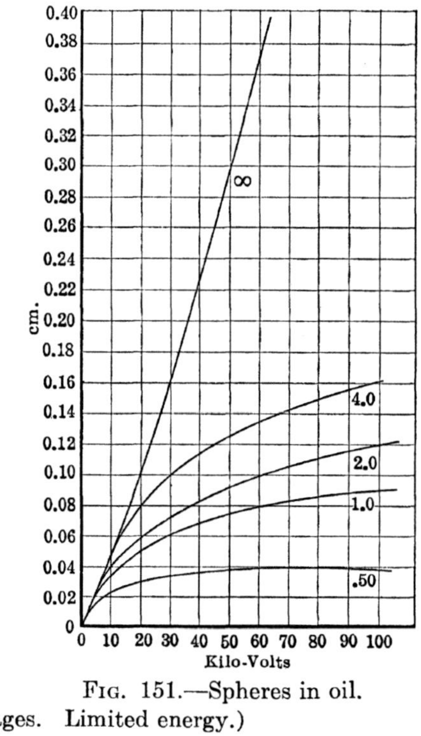

FIG. 151.—Spheres in oil.

(Transient voltages. Limited energy.)

however, does not give a circuit for which the constants can be definitely calculated. Oil is an excellent insulation in combination with barriers. On solid insulations the effect of transient voltages is cumulative. A partial break occurs which is enlarged by each succeeding impulse, until finally dynamic follows. With oil, such "cracks" are closed up by new oil immediately.

One investigation shows that when the energy is limited a voltage is reached where the spark distance becomes a function of the energy and practically independent of the voltage.[1] Figs. 148, 149, 150, 151 taken from this investigation show the spark-over curves for oil and air. The figures on the curves represent the current allowed to enter in the impulse, and thus indirectly indicate, to a certain extent, the energy. The impulse was produced by suddenly applying a continuous voltage to the low side of a transformer (Fig. 147*b*). The curve marked ∞ is for infinite energy supply (60 cycles). If sphere gaps are set in air and in oil for 30 kv. at 60 cycles the spacings are 0.95 cm. and 0.16 cm. respectively. If a transient voltage of 100 kv. is applied, the gap in oil will not spark over if the energy applied to the circuit is less than 1.04 joules (4 amp.). For air the minimum energy is 0.45 joule (1.5 amp.).

At continuously applied high frequency oil breaks down at lower voltages than at 60 cycles. The following comparison made by the author is of interest.

TRANSIL OIL—BETWEEN FLAT TERMINALS—SQUARE EDGE—2.5-CM. DIAMETER—0.25-CM. SPACE—BREAKDOWN VOLTAGE GRADIENTS

60 cycles Kv./mm. (max.)	High frequency alternator 90,000 cycles Kv./mm. (max.)	Single impulse, sine shape, corresponding to 200,000 cycles Kv./mm. (max.)
17	6.7	39

[1] Hayden and Steinmetz, Transactions A.I.E.E., June, 1910.

CHAPTER VII

SOLID INSULATIONS

General.—Some of the principal solid insulations are varnished cambric, oiled and varnished pressboard, built up pressboard, treated wood, mica, micanite, soft and hard rubber, synthetic resins, glass, and porcelain.

With gaseous and liquid insulations as air and pure oil there is very little loss up to the breakdown gradient. A gradient just under the breakdown gradient may be applied and held and the loss is so small that no appreciable heating results. Thus, the loss in air and oil is essentially a phenomenon above the electric elastic limit. This loss generally exists in some locally broken down part of the insulation, as the corona on the surface of a wire. The break does not extend through the whole insulation, and when the stress is removed, new air takes the place of the broken down air and all evidence of overstress is removed; in other words, in air and oil a local breakdown is "self healing."

Almost all insulations are partially conducting or have a very high resistance which is spoken of as insulation resistance. The actual resistance of the insulation itself, which is very high, apparently has no direct connection with the dielectric strength, which is measured by the gradient or flux density or stress required to electrically strain the dielectric above the "electrical elastic limit" so that actual rupture or breakdown occurs. For instance, in a condenser made of two metal plates with a solid dielectric between them, when a.c. potential is applied, energy is stored in the dielectric by electric displacement at increasing potential and delivered back to the circuit at decreasing potential, as long as the potential does not stress the insulation above the elastic limit. If the dielectric were perfect a wattmeter in the circuit would indicate no loss. In all practical insulations the wattmeter does read a loss due to the I^2R loss in the insulation, and the dielectric loss, sometimes called dielectric hysteresis.[1] If the voltage is sufficiently increased to exceed the elastic limit actual rupture or breakdown occurs; along this discharge path the insulation is destroyed. Air has a very high insulation

[1] In what follows the total loss will be called the dielectric loss. See pages 36, 37.

resistance, but not a very high dielectric strength. When the insulation resistance, however, of a given solid insulation becomes very low, as caused by moisture, etc., it is an indication of large loss and low breakdown voltage. Thus, the term "insulation resistance" generally takes into account the resistance of the occluded moisture.

Insulation Resistance.—The actual resistance of the insulating material itself is generally very high. Practically all solid and liquid insulations absorb moisture to a greater or less extent. The capillary tubes and microscopic interstices, etc., in the structure become filled with moisture and gases. In the non-homogeneous structure this makes a complicated arrangement of capacities and resistances in series and in multiple. A simplified diagrammatic illustration of the distribution is shown in Fig. 152.

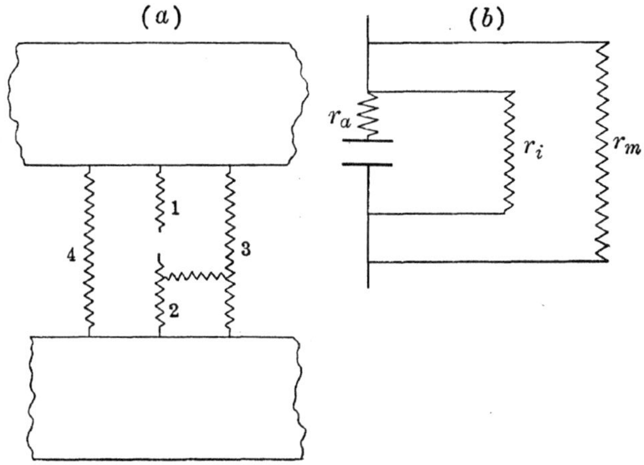

FIG. 152.—Diagrammatic representation of resistance arrangement in imperfect insulations.

Let 152(*a*) represent a magnified section of insulation between two terminals, and 152(*b*) a diagrammatic representation. The resistance of the insulation itself, which is very high, may be represented by r_i. 1 may be a moist fiber which extends only partially through the insulation and is thus in series with a capacity. It may be represented by r_a. 3 and 4 may be wet fibers which extend all the way across, and are represented by r_m. There may also be leakage resistance over the surface.

Direct current is used to measure insulation resistance, as the charging current for a.c. is very large and masks the resistance current. Watt measurements are necessary, as well as volts and amperes, to determine the effective a.c. resistance. The a.c. and d.c. resistance should, however, be quite different. When d.c. is used sufficient time must elapse after the application of voltage to allow for absorption (see Chapter II, page 36).

With d.c. the only path for the current is through r_i and r_m in multiple, which is, therefore, the resistance measured. The resistance varies with the applied potential, decreasing with increasing potential. The conducting particles are caused to line up, cohere, occluded gases break-down, etc., as the potential is increased.

When a.c. voltage is applied to insulation the capacity current must pass through r_a; in shunt with this is the circuit through r_i and r_m. The loss in r_a must increase with increasing frequency, while the loss through r_i and r_m must remain constant at a given voltage, independent of the frequency. The greater loss will generally occur in r_a. The d.c. insulation resistance cannot be used in approximating the I^2r a.c. loss. If the a.c. loss is measured, as well as the voltage and current, the effective resistance may be calculated. This resistance loss, however, must be greater than that due to r_a as other losses are included.

In Table LXX some d.c. resistances are given for different materials. Note the effect of moisture absorbed from the air even for varnished materials.

Table LXX.—Insulation Resistance

Resistance Megohms per Cm. Cube

(Data obtained by Evershed, J.I.E.E., Dec. 15, 1913)

Material	D.c. volts			
	50	100	200	500
Cotton, dry	350.0	275.0	200.0	140
Cotton, not dried	2.8	2.5	2.2	....
Micanite, exposed to air	220.0	175.0	160.0	140
Micanite, dried	....	....	300,000	200,000
Cylinder oil, trace of moisture	....	22,000	22,000	22,000
Cylinder oil, dry	....	36,000	36,000	36,000
Varnished cloth, 12 hr. after baking.	....	35,000	35,000	35,000
Varnished cloth, 9 days after baking.	17,000	14,000	11,000	9,000

With air and oil an appreciable loss begins only when a definite gradient, sufficiently large to cause local rupture as brush or corona, somewhere results; loss occurs after the elastic limit has been exceeded. The dielectric loss in solid dielectrics is essentially a loss below the elastic limit. In solid dielectrics a stress may be applied lower than the elastic limit, which after a short time—on account of the heating and hence weakening of the insulation—will cause rupture.

Rupturing Gradients.—Apply voltage lower than the puncture voltage between concentric cylinders with dry insulation between surfaces, and gradually increase to the rupturing voltage within a short time so that there can be no appreciable rise in temperature. The mechanism of rupture will be quite different for oil or air, glass or porcelain, and cambric. For oil and air corona results near the surface of the inner cylinder, the breakdown is local, and disappears

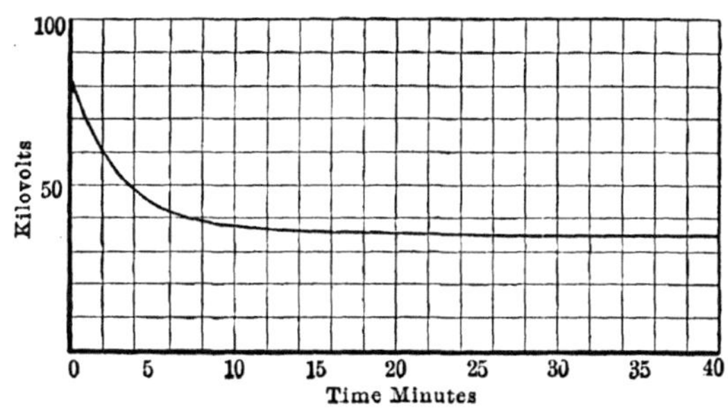

FIG. 153.—Insulation puncture voltage *vs.* time.
(Oiled pressboard, 2.5 mm. thick. The curve does not cut the kv axes as it appears to on account of the time scale.)

when the stress is removed; or, if the stress is further increased spark-over finally results. For glass or porcelain, as soon as overstress is reached locally, a crack results at the surface of the inner cylinder and complete breakdown follows. With cambric there is local rupture and local charring of the insulation, which forms carbon "needles." The break is progressively increased; this continues until the breakdown is complete. A wet thread or gas bubble, as (1) in Fig. 152(*a*), may, in effect, act as a needle and thus cause breakdown. The breakdown gradients of solid insulations are thus variable and not as definite as with air and oil.

The puncture tests on solid insulations vary greatly between different samples of the same material, shape and area of the electrodes, time of application of voltage, etc. Insulations should be thoroughly dried before tests are made.

In comparing solid insulations it is generally best to make some arbitrary time tests to include the effect of dielectric loss, and thus, heating on the breakdown voltage. The effect of loss is cumulative; the insulation becomes warm and while the loss increases with the temperature, the dielectric strength generally decreases with increasing temperature. The ultimate strength naturally, then, depends on the rate at which this heat is conducted away. This is illustrated in Fig. 153. With the electrode used, if the voltage is "rapidly applied" before heating occurs, 80 kv. are required to cause rupture.[1] If 50 kv. are applied rupture does not occur until 4 minutes have elapsed, while 30 kv. may be applied indefinitely without rupture if the room temperature is not increased. The curve does not cut the voltage axes as appears in the cut, on account of the time scale, but is an asymptote to it. The effect of temperature is also illustrated in Table LXXI, where, in one case, the electrode is of brass giving good heat conduction, and in the other case of wood coated with tin-foil giving poor heat conduction. As the applied voltage approaches the "rapidly applied" rupturing voltage there is not sufficient time for the insulation to heat to a great extent and the effect is about the same for either brass or the coated wood electrode.

Table LXXI.—Effect of Heat Conducting Properties of Terminals on Time of Breakdown

Two Thicknesses of No. 12 Oiled Cloth
(From Rayner, Journal I.E.E., Feb. 8, 1912)

Volts	Time of breakdown	
	Brass terminals	Wood terminals[2]
9,000	570.0 sec.	50.0 sec.
10,000	48.0 sec.	19.0 sec.
11,000	16.5 sec.	10.0 sec.
12,000	10.2 sec.	6.2 sec.
14,000	5.2 sec.	4.5 sec.

[1] Rapidly applied voltage as used above means voltage applied within a fairly short time, a few seconds, and not impulse voltage or voltage of very steep wave front. "Instantaneous" is commonly used to designate this test; this term is confusing. The test itself is not wholly satisfactory as it is indefinite. It offers, however, a means of making a preliminary comparison of insulations.

[2] Coated with tin-foil.

The effect of applying a high voltage, allowing different periods of rest for cooling, and then applying voltage until rupture, is shown in Table LXXII. The injurious effects of applying high voltage for different lengths of time is shown in Table LXXIII. This is probably due to the effect of heat and injury to the surface by corona.

TABLE LXXII.—RECOVERY IN VARYING PERIODS OF REST AFTER APPLICATION OF 9000 VOLTS FOR 1 MINUTE[1]

Two Thicknesses of No. 12 Oiled Cloth

(From Raynor, Journal I.E.E., Feb. 8, 1912)

Period of rest	Time to break at 11,000 volts	Period of rest	Time to break at 11,000 volts
0	2.6 sec.	0	2.5 sec.
1 min	9.5 sec.	5 sec.	4.4 sec.
2 min	11.9 sec.	15 sec.	9.0 sec.
Fresh material	12.0 sec.	60 sec.	9.8 sec.
....	....	Fresh material	20.5 sec.

TABLE LXXIII

(From Raynor, Journal I.E.E., Feb. 8, 1912)

Time of treatment at 5000 volts	Time of breakdown (left to cool over night)	
	At 6500 volts	At 7000 volts
0	22.0 min.	8.5 min.
1 hr	6.5 min.	8.3 sec.
2 hr	3.0 sec.	....

TABLE LXXIV.—EFFECT OF AIR GAP (CORONA)

Two Thicknesses of No. 12 Oiled Cloth

10,000 Volts

(Raynor, Journal I.E.E., Feb. 8, 1912)

Air gap, mm.	Time to puncture, seconds
0.00	42.0
0.30	34.0
0.50	27.5
0.75	24.0
1.05	120 (irregular)

[1] Time to break at 9000 volts, 1–½ to 2 minutes.

The action of corona or breakdown of the air at the surface of the insulation is shown in Table LXXIV.

The arbitrary practical tests for *comparing* insulations are the Rapidly Applied (Instantaneous) Test, the One-minute Test, and the Endurance Test. Rapidly applied breakdown voltage is found by applying a fairly low voltage and rapidly increasing until breakdown occurs. Voltage should be increased at about 5 kv. per sec.

The Minute Test is made by applying 40 per cent. of the rapidly applied voltage, and increasing this voltage 10 per cent. at 1-minute periods until puncture occurs. (Total time about 5 min.)

The Endurance Test is made by applying 40 per cent. of the minute test voltage and increasing the voltage 10 per cent. every hour or half hour until puncture occurs. These tests may be made at any given temperature. The electrode should be of a given size and weight. Ten-centimeter diameter electrodes, *slightly* rounded at the edges, will be found convenient. Table LXXV gives an example of such tests.

TABLE LXXV

Muslin with Three Coats Varnish—Total Thickness 2 mm.
Rapidly Applied (Instantaneous) Breakdown and Resistance

Temp., deg. cent.	Resistance in megohms for samples				
	1	2	3	4	5
20	73,400	73,400	97,900	97,900	73,400
75	650	390	310	270	170
100	100	90	73	70	50
	Breakdowns in kilovolts				
100	40.6	42	44	43	36.5

Average breakdown 40.6 kilovolts

One-minute Test

Apply 40 Per Cent. of Rapidly Applied Breakdown Voltage for 1 Minute, with 10 Per Cent. Increase in Voltage Each Minute

Temp., deg. cent.	Resistance in megohms				
	6	7	8	9	10
20	73,400	73,400	73,400	97,900	97,900
75	320	333	451	274	330
100	50	60	70	40	60

Time under stress	Potential applied	Temperature for samples					
		6	7	8	9	10	Air
Start	....	100	100	101	103	100	100
1 min.	16,000	101	101	103	104	102	100
2 min.	17,500	102	102	104	105	103	100
3 min.	19,000	105	104	106	108	105	100
4 min.	20,500	111	108	108	111	107	100
5 min.	22,000	121	112	111	117	108	100
5 min. 5 sec.	23,500	......	......	....	....	115 punctured	
5 min. 46 sec.	23,500	146 punctured		....	....	....	....
6 min.	23,500	......	118	114	129	....	100
6 min. 10 sec.	25,000	......	......	....	135 punctured		....
6 min. 35 sec.	25,000	......	......	118 punctured		....	....
7 min.	....	......	135	....	....	....	100
7 min. 58 sec.	26,500	......	140 punctured		......	....	....

Endurance Test

Apply 40 Per Cent. of 1-minute Period Endurance Voltage, for 30-minute Periods Endurance Test, with 10 Per Cent. Raise in Voltage Each Period

Temp., deg. cent.	Resistance in megohms				
	11	12	13	14	15
22	74,300	74,400	74,400	99,100	49500
75	390	420	230	290	110
100	90	70	30	60	30

Time under stress	Potential applied	Temperature					
		11	12	13	14	15	Air
Start	....	99	100	100	101	99	100
0.5 hr.	10,000	103	107	133	108	115	100
34 min.	11,000	......	......	182 punctured		......	....
54 min.	11,000	......	......	......	....	198 punctured	
1 hr.	11,000	104	108	......	112	....	100
1.5 hr.	12,000	106	111	......	114	....	100
2 hr.	13,000	106	118	......	122	....	100
2 hr. 23 min.	14,000	......	......	......	215 punctured		....
2 hr. 25 min.	14,000	......	203 punctured		....	....	....
2.5 hr.	14,000	136	......	......	....	....	....
3 hr.	15,000	143	......	......	....	....	....
3 hr. 2 min.	16,000	145 punctured		...	....	....	....

A considerable amount of data is given here for the 1-minute time test. It must be remembered that in design only a fraction of the maximum gradient corresponding to this voltage is permissible for continuous operation, the particular per cent. depending upon the design, the insulation, the rapidity at which heat may be radiated, or conducted away, etc. It is generally not more than 10 per cent.; it is often as low as 5 per cent., sometimes as high as 30 per cent.

Insulation tests are generally made for convenience on sheets between flat terminals. The gradient at the edges, even when these are rounded, is generally higher than the average gradient e/x. This edge effect is different with different thicknesses of insulation. The puncture voltage per centimeter thickness is always greater for thin sheets of insulation than for thick ones. This is partly due to the edge effect which cannot be corrected for and becomes relatively greater as the thickness of insulation is increased. (At small spacings e/x is very nearly the true gradient; at large spacings e/x is not the true maximum gradient.)[1] It is also greatly due, in the time tests, to the better heat distribution and dissipation in the thin sheets and partly due to the fact that energy is necessary for disruption, so that when the rupturing distance is limited, as in the case of thin sheets, the apparent strength increases. Fig. 154 shows apparent variation in strength with thickness. Between parallel plates the apparent strength is approximately:

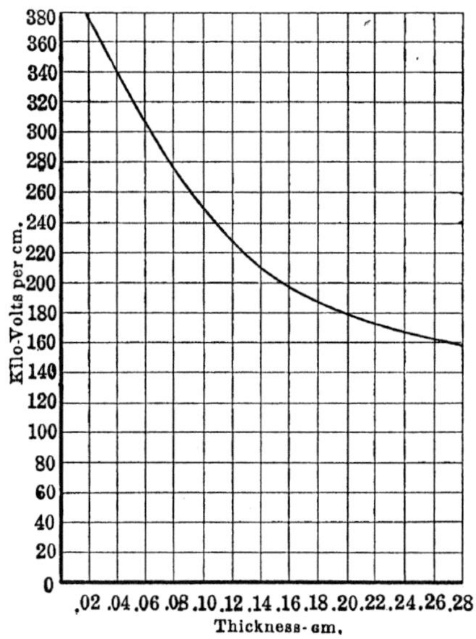

FIG. 154.—Insulation strength *vs.* thickness.

$$g_s = g\left(1 + \frac{a}{\sqrt{t}}\right)$$

where g and a are constants and t is thickness.

When shielded edges are used, it is generally best to make insulation tests under oil unless it is desired to make a study of the

[1] Special flat terminals are sometimes used with "shielded" edges so that e/x is the true gradient.

effect of corona on the insulation. With pointed electrodes the instantaneous puncture voltage of a given insulation will be less in oil than in air. This is not because the oil weakens the insulation but because corona forms on the point in air and spreads over the surface giving the effect of a flat plate electrode.

Solid and Laminated Insulation.—The structure of most insulations is not homogeneous. If a given insulation is tested with terminals of varying area it is found that the average puncture voltage becomes lower as the area is increased, and thus the chance of it covering a weak spot is increased. As would be expected this approximately follows the probability law as shown in Fig. 155.

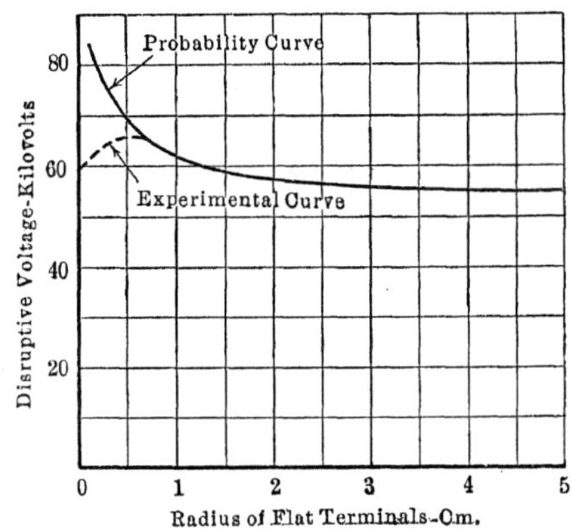

FIG. 155.—Insulation strength *vs.* area of terminal.

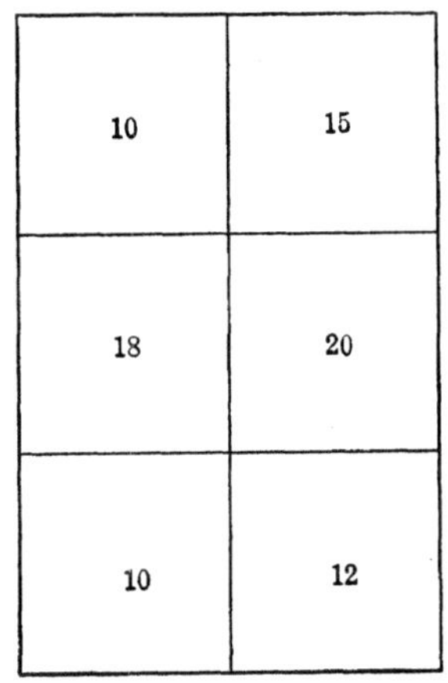

FIG. 156.—Variation in dielectric strength at different parts of a piece of insulation.

The reason that the experimental curve bends down is that the gradient is fairly constant until the terminal becomes small when the gradient increases, due to flux concentration with the smaller terminal. If the curves were plotted with actual gradients instead of voltages, the experimental curve would follow more closely the probability curve. The experimental gradient curve would, however, for the smaller sizes bend up faster than the probability curve, due to greater apparent strength for smaller terminals, as for small spheres in the case of air. The curve between e/x and diameter for flat terminals of the larger sizes when the concentration at the edges is about constant should follow the probability curve.

For example, suppose Fig. 156 represents a piece of solid insulation 0.025 mm. thick and sufficiently large to contain every condition of "weak spot." Divide this into six equal squares each of area *a*. The strength is marked on the various areas. Assume that an electrode is used giving no edge effect. With electrode of area *a* six tests are required to go over the whole piece. With electrodes of area 2*a* three tests are required, with area 3*a* two tests, and 6*a* one test. The following results may be obtained:

Area electrode	No. of punctures	Total area covered	Volts per mm.		
			Maximum	Minimum	Average
a..........	6	6*a*	20	10	14
2*a*..........	3	6*a*	18	10	13
3*a*..........	2	6*a*	12	10	11
6*a*..........	1	6*a*	10	10	10

The results are somewhat similar to the lower points of the curve in Fig. 155.

On account of these characteristics alone an insulation built up of laminations is much better than a solid insulation as the weak spots in the laminations are not likely to line up. It is also much easier to make better and more uniform insulation in thin sheets. Another probable important reason for greater strength of laminated insulation is that the energy distances are interrupted by the discontinuous surfaces.

Tests are useless for comparing insulation strengths unless made upon some standard basis. Great caution is necessary in the use of tabulated values of insulation strength in design. On account of the variable quality of solid insulation, tests must be continually made to see that the product does not change. Vacuum treatment is necessary before use to remove moisture. Even when all of the test conditions are known, experience is necessary to judge the proper factor of safety. Aside from this, stress concentrations due to the shapes and spacings of the conductors must always be considered and allowed for. It is generally not possible to do this with mathematical exactitude, but approximation must be made with all factors in mind. Care must be taken that the solid insulation is below the rupturing gradient at any local point. If such a point is broken down locally the flux becomes still further concentrated. The puncture voltage will decrease with frequency, even over the commercial range, due to increasing loss with increasing frequency.

Impulse Voltages and High Frequency.—It takes energy and therefore time to rupture insulation. For a given potential a given number of cycles of very high frequency voltages, where heating does not result, are therefore much less injurious than the same number of cycles at low frequency. This also applies to impulse voltages of steep wave front. Continuously applied high frequency is, however, generally very injurious for two distinct reasons:

(1) On account of the very great loss at high frequency the insulation may be literally burned up in a very short time even at low voltages. This condition does not result in practice from surges, etc., on low frequency lines, but in high frequency generators, and transformers, etc. In such apparatus it is important to use very smooth electrodes to prevent local concentration of stress and charring of insulation. This is especially so where contact is made with the air. If a local brush starts, on account of the great loss, it becomes very hot and extends out a considerable distance.

(2) In certain apparatus containing inductance and capacity very high local potential differences may be produced by resonance and thus cause rupture by overpotential. The high frequency thus does not cause the rupture directly but makes it possible by causing overpotential. Local concentration of stress may also result in non-homogeneous insulation, as across the condenser and resistance combinations in Fig. 152.

The term "high frequency" is generally used in such a way that no distinction is made between sinusoidal high frequency from an alternator, undamped oscillations, damped oscillations, impulses of steep wave front, etc. Naturally the effect of continuously applied undamped oscillations is quite different from a single high-voltage impulse of extremely short duration. As the effects are attributed to the same cause—"high frequency"—apparent discrepancies must result. (See comparative tests, Table LXXVI, page 184.)

If the time of application is limited below a definite value, higher voltages are necessary to produce the same results in the limited time. Impulse voltages of steep wave front many times in excess of the rupturing voltage may be applied to insulations without rupture if the application is very short—measured in microseconds. They may be caused in practice by lightning, switching, etc. If such voltages are sufficiently high, complete rupture may result at once. In any case if these voltages are higher than the 60-cycle puncture voltages the insulation will be damaged. As an example, an impulse voltage equal to three times the 60-cycle puncture voltage may be applied to a line insulator. During the very small time between the application of the voltage and the arc-over through the air, the insulator is under great stress. It may be that up to the ninth application of such a voltage there is no evidence of any injury, while on the tenth application failure results. Each stroke has contributed toward puncture. It is probable that each application adds to or extends local cracks.

Cumulative Effect of Overvoltages of Steep Wave Front.—Voltages greatly in excess of the "rapidly applied" 60-cycle puncture voltage may be applied to insulation without rupture if the time of application is sufficiently short. All such overvoltages injure the insulation, probably by mechanical tearing, and the effect is cumulative. A sufficient number will cause breakdown. For example: A piece of oiled pressboard 3.2 mm. thick has a rapidly applied breakdown at 60 cycles of 100 kv. maximum. If sinusoidal impulses reaching their maximum in 2.5 microseconds are applied, the number of impulses to cause break-down is as follows:

Kv. maximum of impulse applied	Number to cause breakdown
100	∞
140	100
150	16
155	2
165	1

If the impulses are of still shorter duration, a greater number are required to cause breakdown at a given voltage. Insulations, and line insulators, are often injured and gradually destroyed in this way by lightning.

Strength vs. Time of Application.—It was stated above that the strength varies with the time of application. A curve is given in Fig. 153. The range of time shown in this curve is from a few seconds ("instantaneous") to an indefinitely long time. Over the greater part of the plotted curve heating is a factor and the great decrease is principally due to heating. Where the time of application is much less than "instantaneous" value, and heating can have no appreciable effect, the strength still increases very rapidly as the time of application is decreased. The increased strength at this part of the curve is due to limited energy, as explained on page 177. Some values from an actually measured curve are:

(14 Layers Impregnated Paper between Concentric Cylinders)

$$R = 0.67 \text{ cm.}$$
$$r = 0.365 \text{ cm.}$$

Time, sec.	Kilovolts to puncture (maximum)
∞	32.6
60.0	37.5
1.0	49.3
0.1	61.0
0.01	85.0
0.001	113.0
0.0001	196.0
0.00001	550.0 (calculated)

$$g = 15.5\left(1 + \frac{0.5}{\sqrt[4]{T}}\right) \text{ kv./mm. max.}$$

The equation of the strength-time curve over the complete range obtained from an examination of a number of curves is of the form

$$g = g_s\left(1 + \frac{a}{\sqrt[4]{T}}\right)$$

when T is the time of application in seconds, and g_s is the gradient in kv./m.m. for indefinite time.

Both a and g_s vary with the thickness of the insulation, temperature, etc.

In order that the strength may be high for indefinite time, the loss should be low.

Permittivity of Insulating Material.—In design, a knowledge of the permittivity of insulating materials is as important as the dielectric strength. For solid insulations the permittivity increases with the specific gravities of the material in almost a direct ratio. The various properties of some of the common insulations are given in Table LXXVI.

TABLE LXXVI.—DIELECTRIC STRENGTH OF SOLID INSULATIONS

Variation of Dielectric Strength with Thickness and Number of Layers. (One-minute Tests—60-cycle—10-cm. Terminals in Oil. Values Effective Sine Wave)

PRESSBOARD

No. of layers	Thickness per layer, mm.	Total thickness, mm.	Kv./mm.—temperature 25 deg. C.		
			Varnished, kv./mm.	Oiled transil, kv./mm.	Linseed oil, kv./mm.
1	0.178	0.178	25.3	....	25.3
1	0.254	0.254	26.3	39.3	23.6
1	0.508	0.508	16.1	....	17.7
1	0.787	0.787	19.1	28.0	23.5
1	1.575	1.575	15.5	29.2	19.0
1	2.390	2.390	11.1	23.0	15.1
1	3.170	3.170	9.5	21.1	14.2
2	0.178	0.356	22.5	....	21.0
2	0.254	0.508	17.7	33.5	19.7
2	0.508	1.016	12.8	....	17.2
2	0.787	1.574	13.7	22.9	20.2
2	1.575	3.150	10.6	19.7	15.3
2	2.390	4.790	7.54	16.1	12.1
2	3.170	6.340	6.3	14.7	11.0
4	0.178	0.712	26.7	....	16.6
4	0.254	1.016	22.6	29.8	15.3
4	0.508	2.032	16.8	....	15.8
4	0.787	3.148	16.5	20.6	19.7
4	1.575	6.300	11.6	13.65	12.7
4	2.390	9.560	9.4	11.5	....
4	3.170	12.680	6.6	....	....
6	0.178	1.068	25.5	....	15.0
6	0.254	1.524	20.6	27.5	13.1
6	0.508	3.048	15.4	....	14.4
6	0.787	4.702	14.9	20.0	17.8
6	1.575	9.450	....	11.3	....

TREATED WOOD

Temperature 25 deg. C.

	Across grain	kv./mm.		With grain	kv./mm.
1	12	6.42	1	30	2.47
1	15	4.53	1	60	1.57
1	20	3.85	1	90	1.27
1	25	3.02	1	120	1.12

Paper
(Untreated)

No. of layers	Thickness per layer, mm.	Total thickness, mm.	25° C., kv./mm.	100° C., kv./mm.
1	0.064	0.064	9.3	9.3
1	0.127	0.127	8.7	7.9
1	0.254	0.254	7.9	7.3
4	0.064	0.258	8.7	8.3
4	0.127	0.508	7.5	6.7
4	0.254	1.016	6.6	6.2
8	0.064	0.516	8.7	8.1
8	0.127	1.016	7.4	6.6
8	0.254	2.032	6.3	6.0

Varnished Cloth

No. of layers	Thickness per layer, mm.	Total thickness, mm.	25° C., kv./mm.	100° C., kv./mm.
1	0.305	0.30	26.2	23.6
2	0.305	0.61	20.5	19.7
3	0.305	0.91	18.5	17.0
4	0.305	1.22	16.8	14.9
5	0.305	1.52	15.5	13.1
6	0.305	1.83	14.6	11.5
7	0.305	2.13	14.0	10.3
8	0.305	2.44	13.3	9.2
9	0.305	2.74	12.8	8.3
10	0.305	3.05	12.3	7.5

Hard Rubber

No. of layers	Thickness per layer, mm.	Total thickness, mm.	Puncture voltage	
			25° C., kv./mm.	100° C., kv./mm.
1	0.0193	....	59.7	....
1	0.0223	....	55.6	....
1	0.0312	....	48.7	....
1	2.0	2.0	17.3	....
1	3.0	3.0	14.2	....
1	4.0	4.0	12.7	....
1	5.0	5.0	11.8	....
1	6.0	6.0	11.2	....

MICA

1	0.0508	0.0508	96.5	....
1	0.1016	0.1016	88.7	....
1	0.1524	0.1524	75.5	....
1	0.2032	0.2032	62.6	....
1	0.5080	0.5080	26.1	....

GLASS

Total thickness	Kv./mm.
6.0	10

The strength of glass decreases rapidly with thickness.

PORCELAIN

Total thickness, mm.	Kv./mm.
0.5	16.0
1.0	14.5
2.0	12.2
5.0	11.0
10.0	9.6
15.0	9.2

VARIATION OF INSULATION STRENGTH WITH TIME OF APPLIED VOLTAGE

Material	Thickness, mm.	Time to puncture, min.	25° C. kv./mm.	100° C. kv./mm.
Oil impregnated paper, 30 layers, 60-cy., 10-cm. diameter discs, round edges.	1.90	"Inst."	39.4	32.0
		1	33.1	27.3
		2	31.0	25.7
		4	29.2	24.5
		6	28.2	23.6
		10	26.8	22.7
		20	25.5	21.6
		40	23.6	20.5
		60	22.7	19.7
		80	22.1	19.3
		100	21.6	19.0

Note that for a given thickness the strengths of materials do not vary as greatly as might be expected.

Material	Thickness, mm.	Time to puncture, min.	Puncture voltage, 25° C.		
			1 layer, 0.30 mm. kv./mm.	5 layers, 1.50 mm. kv./mm.	10 layers, 3.0 mm. kv./mm.
Varnished cloth, 0.30 mm., 60-cy., 10-cm. diameter in air, round edges	0.30	"Inst."	52.5	36.1	27.2
		0.1	37.7	27.5	22.3
		0.2	34.5	25.8	19.7
		0.5	32.8	22.9	16.4
		1.0	32.5	21.0	14.3
		2.0	32.1	20.3	13.0
		3.0	31.8	19.9	12.6
		5.0	31.5	19.8	12.5
		10.0	31.1	19.7	12.3
		20.0	30.2	19.6	9.9

PERMITTIVITY OF INSULATING MATERIALS

	Permittivity	Specific gravity
Asphalt	2–½	
Bakelite	4–½	
Cambric (varnished)	4–½ to 5–½	1.15
Fiber (horn) dry	2–½	0.7 to 1
Fiber (horn) oil	4–½ to 5	0.9 to 1.5
Glass (crown)	6	3 to 3.5
Glass (heavy flint)	10	4.5
Gutta percha	3–½ to 4	
Lead stearate	5.2	
Lead palmitate	5.2	
Lead oleate	5	
Mica	5 to 7	
Oil (linseed)	3–½	0.95
Oil (transil)	2 to 2–½	0.8 to 0.9
Paper (dry)	2.6	1.00
Paper (paraffined)	3–½	
Paper (oiled)	4	1.25
Paraffine	2 to 2.3	0.9

Permittivity of Insulating Materials.—*Continued*

	Permittivity	Specific gravity
Pressboard (dry)	3	1.25
Pressboard (oiled)	4 to 6	1.40
Porcelain	4-½ to 5	2.4
Rubber (hard)	3	
Rubber (vulcanized)	2-½	
Shellac	3	
Sulphur	4	
Wood (treated)	3 to 3-½	0.8 to 0.9

Comparative Insulation Strength for High Frequency, Impulse, Oscillation and 60-cycle Voltages

Temperature 30 deg. C.

60 Cycles		High frequency (alternator), 90,000 cycles		Damped oscillation, train freq. 120 sec., 200,000 cycles		Single impulse, sine shape, corresponding to half cycle of 200,000 cycle	Thickness, mm.	Layers
Inst.	1 min.	Inst.	1 min.	Inst.	1 min.			
kv./mm. (max.)		kv./mm. (max.)		kv./mm. (max.)		kv./mm. (max.)		
Transil Oil between Flat Terminals—Square Edge. 2.5-cm. Diameter, 0.25-cm. Space								
17		6.7		30		39		
Oiled Pressboard. 10-cm. Diameter—Square-edge Discs in Oil								
35.5	31.0	9.5	7.3	37.0	29.0	72.0	2.5	1
39.5	37.0	6.1	4.1	42.0	24.0	....	5.0	2
......	....	2.5	1.76	....	....	....	15.0	3
Varnished Cambric								
53.0	46.5	19.5	17.6	....	....	108.0	0.6	2
42.0	31.0	13.5	10.0	55.0	56.0	78.0	1.5	5
42.0	31.0	10.0	7.3	49.0	41.0	70.0	2.5	8
33.0	27.5	....	....	41.0	30.5	60.0	3.6	12

Energy Loss in Solid Insulation.—In general, energy loss in solid insulation:

(1) Increases with increasing voltage.
(2) Increases with increasing temperature.
(3) Increases with increasing frequency.
(4) Increases with increasing moisture content.
(5) Increases with increasing impurities, as occluded air, etc.

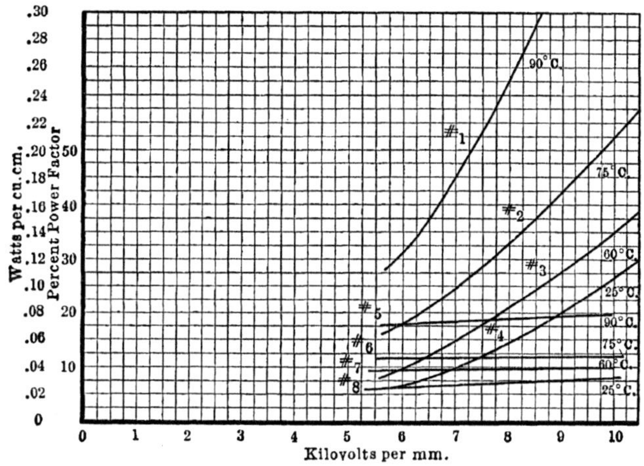

FIG. 157.—Insulation loss and power factor.
(Oiled pressboard 5.0 mm. thick between parallel plates with rounded edges in oil. Curves 1, 2, 3, 4, watts per cu. cm. Curves 5, 6, 7, 8, Power factor.)

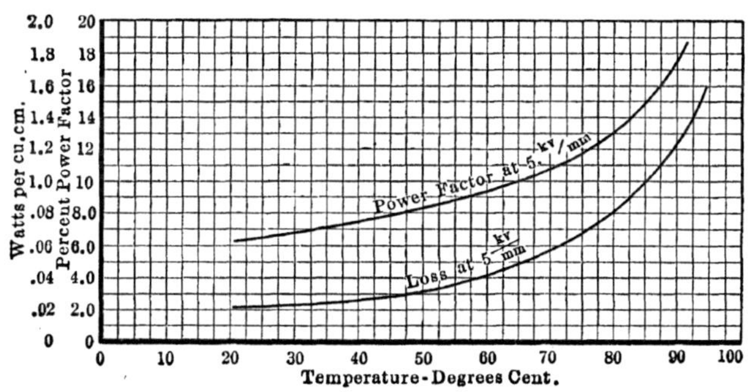

FIG. 158.—Insulation loss and power factor *vs.* temperature.
(Data, Fig. 157.)

For good uniform insulation free from foreign material, moisture, etc., the loss at constant temperature and frequency varies approximately as the square of the applied voltage. The author has found that approximately for good insulations

$$p = afe^2 = bfg^2 \times 10^{-6} \text{ watts/cu. cm.}$$

where f = frequency in cycles per sec.

g = gradient kv./mm.

b = constant varies with different insulations. It is 8 to 10 for varnished cambric in a uniform field, and thickness in order of 5 mm.

Compare calculated values with values given in Table LXXVII at both 60 cycles and high frequency (25 deg. C.). With occluded air or water, where the I^2r loss becomes large in comparison with the "hysteresis" loss, the rate of increase is greater. Due to combinations of resistance and capacity it may then take the form,[1] $p = bfg^2 + (cf^2g^2 + ag^2)$. Fig. 157 shows characteristic curves between energy loss, power factor, and voltage of insulation in good condition.

The loss increases very rapidly with temperature. The temperature curves are shown in Fig. 158. The effect of exposing to the air is also shown. Loss in different insulations is given in Table LXXVII.

TABLE LXXVII.—INSULATION LOSSES

(Effective Sine Wave 60 Cycles)

Total thickness, mm.	Insulation	No. of layers	Temp., deg. C.	Volts per mm.	Watts per cu. cm.
4.0	Varnished cloth	15	25	4,000	0.005
				6,000	0.015
				8,000	0.035
				10,000	0.060
				12,000	0.090
4.0	Varnished cloth	15	90	4,000	0.025
				6,000	0.075
				8,000	0.150
				10,000	0.240
				12,000	0.350
2.5	Oil-treated paper	30	25	10,000	0.040
				14,000	0.070
2.5	Oil-treated paper	30	60	10,000	0.043
				14,000	0.080
2.5	Oil-treated paper	30	90	10,000	0.050
				14,000	0.100
2.5	Oil-treated paper	30	120	10,000	0.050
				14,000	0.100

These losses may be lower or very much higher, depending upon the condition of the insulation.

[1] See Fig. 152.

HIGH FREQUENCY LOSS IN DIFFERENT INSULATIONS
(Alexanderson, Proc. Inst. Radio Engrs., June, 1914)

Volts per mm.	Frequency kilocycle	Material (thickness 5 mm.)							
		Glass		Mica		Varnished cambric		Asbestos	
		Loss watts, cu. cm.	P. F. avg.	Loss watts, cu. cm.	P. F. avg.	Loss watts, cu. cm.	P. F. avg.	Loss watts, cu. cm.	P. F. avg.
500	20	0.02	......	0.02	......	......	......	0.7	......
500	40	0.05	1.4	0.04	1.8	0.08	......	1.3	......
500	60	0.08	......	0.05	......	0.12	3.3	2.0	......
500	80	0.11	......	0.08	......	0.18	......	......	32.0
1000	20	0.08	......	0.07	......	......	......	3.0	......
1000	40	0.20	1.4	0.12	1.8	0.28	......	5.3	......
1000	60	0.32	......	0.24	......	0.45	3.3	7.6	32.0
1000	80	0.40	......	0.35	......	0.65	......	......	......

The rate of increase of the loss with the frequency will vary greatly if the insulation contains moisture. If the moisture is arranged in such a way as to approximate a condenser and resist-

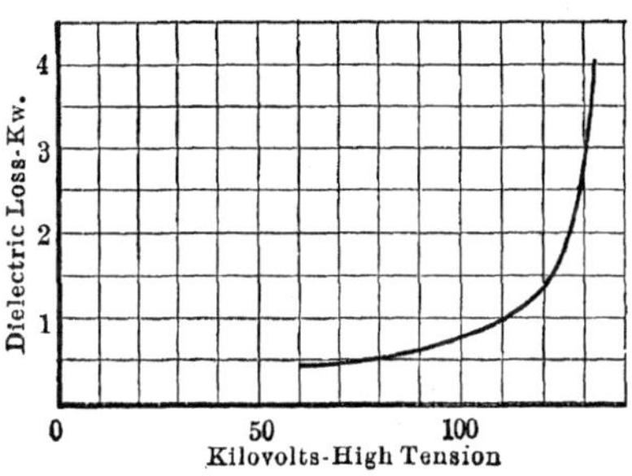

FIG. 159.—Dielectric loss *vs.* voltage in a 2750 kv-a, 120 kv, 60~ transformer.

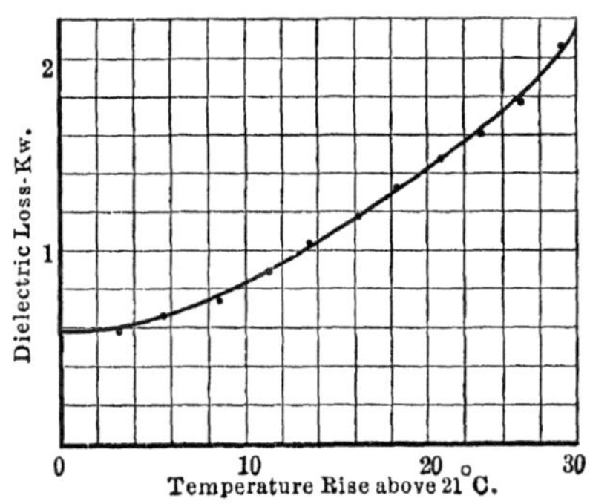

FIG. 160.—Dielectric loss *vs.* temperature in a 3750 kv-a, 120 kv. 60~ transformer at 100 kv.

ance in series, as in Fig. 152, the loss may, over a limited range, increase approximately as the square of the frequency. If the insulation is in good condition the loss may increase approximately directly as the frequency.

The best method of comparing different insulations is by measurement of losses.

Operating Temperatures of Insulations.—The maximum operating temperature of insulations is indefinite. For low-voltage apparatus, temperatures not high enough to cause direct electrical failure may cause mechanical failure in short periods by drying out the insulation, cracking, etc. The maximum safe temperature, at which the life is not greatly shortened mechanically by cracking, drying out, etc., varies with different insulations, but is approximately as follows:

Fibrous materials, cloth, varnish, etc. . . . . 100 deg. C.
Asbestos, mica and similar materials, in combination with binders' varnish, etc. . . . . 150 deg. C.
Mica, asbestos—alone. . . . . very high.

These values are given without consideration of the electrical effects. Often the electrical properties will limit the temperatures below these values. Heat conductivity must be considered in design.

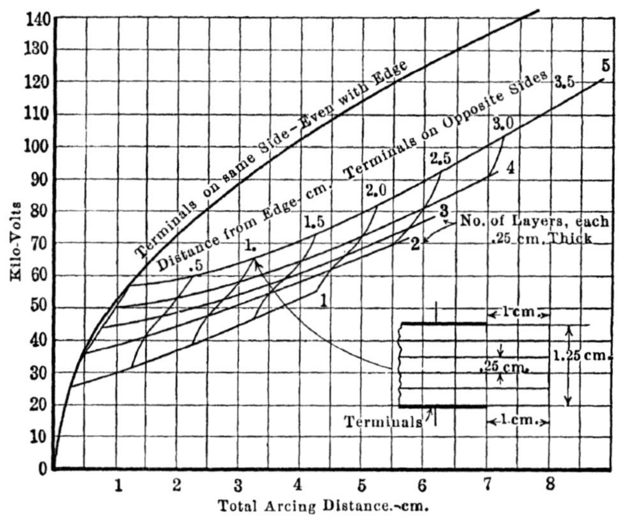

FIG. 161.—Surface arc-over in oil. (Hendricks.)

Surface Leakage.—Strictly speaking, on clean dielectric surfaces, appreciable leakage does not occur. What is generally termed "surface leakage" is a rupture of one dielectric at the surface of another dielectric by overflux concentration due to difference in permittivities, etc. For instance, on a porcelain insulator in air the flux may be sufficiently concentrated at portions of the surface to cause the air to rupture. The *appearance* is that of leakage over the porcelain surface. "Surface leakage" then is quite indefinite, and for a given leakage distance depends upon the position and shape of the electrodes, relative capacities of the materials, etc. (See Fig. 161.) The effects of actual leakage and apparent leakage in air may be seen in Table LXXVIII, page 190. In (1) the spark-over is given with the surface out. In (4) the decrease must be due to actual leakage as there is no flux concentration. In (2) and (3) flux concentration is balanced against increased surface.

Solid Insulation Barriers in Oil.—Properly placed barriers of solid insulation in oil greatly increase the strength of the oil spaces by limiting the thickness of the oil, preventing moisture chains lining up, etc. This should always be considered in design. In placing barriers, however, the arrangement should be such that the stress on the oil is not increased by the higher permittivity of the solid insulation. Data for a simple arrangement are given in Table LXXVIII. It will be noted that in most cases the strength is not greatly increased on account of difference of permittivity of oil and solid insulation. The reliability is much greater, however.

Table LXXVIII.—Pressboard Barriers in Oil

Gap 0.24 cm.	Kv. effective
Oil only	23.0
1—0.08-cm. sheet of pressboard midway between electrodes	24.7
1—0.08-cm. sheet of pressboard against one electrode	22.7

Gap 0.40 cm.	Kv. effective
Oil only	35.3
2—0.08-cm. sheets of pressboard against one electrode	36.5

Gap 0.56 cm.	Kv. effective
Oil only	53.9
3—0.08-cm. sheets of pressboard against one electrode	44.0

Gap 1.16 cm.	Kv. effective
Oil only	96.0
1—0.08-cm. sheet of pressboard 0.1 cm. from each terminal	95.3
1—0.08-cm. sheet of pressboard 0.33 cm. from each terminal	102.0
2—0.08-cm. sheet of pressboard at midpoint	88.5
1—0.08-cm. sheet of pressboard on each terminal	94.5

Strength of pressboard—0.08 cm., 15 kv.; 0.16 cm., 28 kv.

Terminals 6 cm. in diameter, rounded with a 3-cm. radius at the edges.

SURFACE ARC-OVER IN AIR

		Effective kv.	kv./cm.
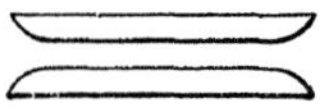	(1) 7.6 cm. Air only between parallel plates. Rounded edges	128.0	16.8
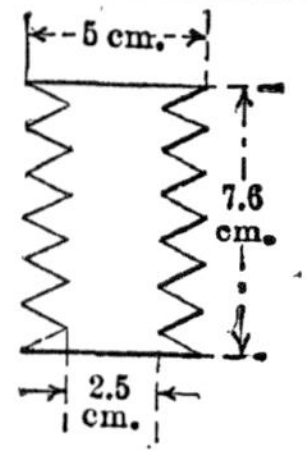	(2) 7.6 cm. Over corrugated rubber cylinder between above plates. Surface oiled	94.0	12.4
	Surface dry	87.0	11.4
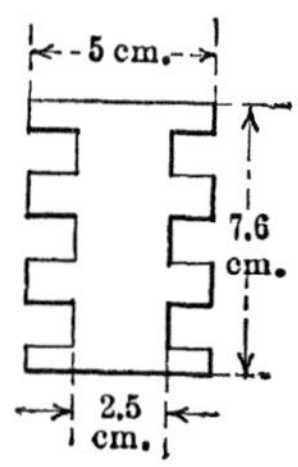	(3) Corrugated rubber cylinder between above plates. Surface oiled	109.0	14.3
	Surface dry	89.0	11.7
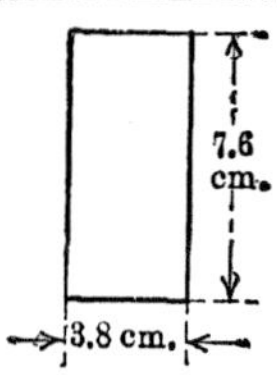	(4) Smooth rubber cylinder between above plates. Oiled surface	122.0	16.0
	Surface dry	62.0	8.0

NOTE.—Maximum possible kv. arc-over 128. Data shows that although corrugations increase stress, actual gain is made by their use by reduction of true leakage.

Impregnation.—Insulations, such as dry paper, with low dielectric strength and low permittivity, are impregnated with oils or compounds of high strength and permittivity. The result is a dielectric of greater strength and permittivity. If the impregnating is improperly done, for instance so as to leave oiled spots and unoiled spots, the dielectric strength may be less than the dry paper alone. This is due to the difference in permittivities of the dry and oiled spots, which causes a concentration of stress on the electrically weak dry spots.

Mechanical.—It is of great importance to arrange designs in such a way that local cracking, or tearing is not caused by high localized mechanical stresses. This is especially so with porcelain, as in the line insulators. Expansion of a metal pin, localized mechanical stress due to sharp corners, expansion of improper cement, etc., will cause gradual cracking of the porcelain. The so-called deterioration of line insulators is often caused in this way.

Direct Current.—As the breakdown depends upon the maximum point of the voltage wave, the direct-current puncture voltage in air and oil is ($\sqrt{2}$) or 1.41 times the sine wave alternating puncture voltage. In air and oil there is very little loss with alternating current until the puncture voltage is reached. With solid insulation losses occur with alternating current as soon as voltage is applied. In the time tests the insulation is thus considerably weakened by heating, especially thick insulation. This causes decreasing breakdown voltages in the time tests as the frequency is increased. In certain insulations the loss must be very small for direct current. The gain for direct current in solid insulation in the time tests is thus greater than the 41% given above.

CHAPTER VIII

THE ELECTRON THEORY

A brief review of the electron theory will be given in this chapter.

A gas is a very poor conductor of electricity. A condenser or electroscope may be left charged in air a great length of time without considerable loss or leakage. If, however, the surrounding air and the terminals of the condenser or electroscope are subjected to the action of X-rays, ultraviolet light, or radio-active substances, the leakage becomes quite rapid. All of these agents in some way act as carriers or change the nature of the gas so that the current passes from terminal to terminal. The gas is said to be ionized.

FIG. 162.—Cathode ray tube.

If terminals are placed in a vacuum tube and high voltage is applied between them, a visible discharge or beam of rays is shot out from the cathode. These cathode rays proceed in straight lines. A pin-hole diaphragm may be placed in their path and a narrow beam obtained (see Fig. 162). This beam may be deflected by a magnetic or dielectric field. J. J. Thomson pointed out that it acts in every way as if it were made up of negatively charged particles traveling at very high velocities.[1] Every test that has been made bears this out. Where the particles strike the glass it becomes luminescent. These particles of negative electricity or "charged" corpuscles are called electrons. The velocity, "charge," and mass of these electrons have been measured.

[1] These rays have been made use of in an oscillograph. In this instrument the beam acts as a pointer and is made to trace a curve under influence of the fields produced by the current or voltage of the wave which is being measured.

A "charged" body in motion is deflected by the electric fields in the same way as a wire-carrying current. The deflection depends upon the ratio of the "charge" e and the mass m. By noting the deflection of the cathode rays in the electric field, J. J. Thomson found the value of the ratio e/m. The most accurate value of the ratio with e measured in electromagnetic units is 1.8×10^7. Each ion in a gas acts as nuclei in the condensation of water vapor. The condensation in the presence of the ions may be made to occur by change of pressure. By observing the rate of fall of the cloud the number of drops or electrons can be calculated. If the total "charge" is measured, e can be at once obtained. This was done by C. T. R. Wilson. e was also later determined by Millikan and found to be 4.77×10^{-10} electrostatic units or 1.6×10^{-20} electromagnetic units. The mass of the electron seems to be about 1/1800 of the hydrogen atom, or the same mass as the hydrogen ion in electrolytic conduction. This mass is about 8.9×10^{-28} grams when the velocity is considerably below that of light, and apparently changes with velocity. It must be considered as that determined by force divided by acceleration. The velocity of the electron varies from 10^7 to 10^9 cm./sec.

The beta particle of radio-active substances is identical with the cathode particle or electron. The gamma rays are, probably, ether waves produced by the action of the beta particles in a way similar to that in which X-rays are produced by the impingement of cathode particles on solids. The alpha particle is a positively charged atom of helium.

The elemental positive particles corresponding to the electron have so far not been found. Ion is a general term used for positive or negative atoms or molecules, electrons, or positive particles.

Take two electrodes in air and apply some low potential between them. Direct ultraviolet light upon the negative electrode. If the voltage is gradually increased, the current increases almost directly up to (*a*), Fig. 163. There is then a considerable range between (*a*) and (*b*) where an increase in voltage does not greatly increase the current. At (*b*) the current suddenly increases very rapidly with increasing voltage. It appears that negative "particles" or electrons are produced or set free at the negative conductor by the ultraviolet light. These "particles of electricity" are attracted to the positive conductor and thus show as current in the galvanometer in the circuit. The number reaching the positive conductor increases with increasing voltage. The current thus increases with increasing voltage. The potential at (*a*) is sufficient to cause practically all of the negative particles that are produced by the light to reach the positive conductor. An increase of the voltage above the value at (*a*) can thus cause very little increase in current unless a new source of ionization is applied or the number of ions is increased in some way. When the voltage is raised above the value at (*b*), the current increases very rapidly with increasing voltage. A new source of ionization has resulted.

The velocity at which ions travel increases with increasing voltage or field intensity. The new source of ionization results when the ions have reached a definite velocity in their mean free path. Townsend has explained this on the hypothesis that ions traveling at sufficient speed generate new ions on collision with neutral atoms or molecules. Thus after a sufficient voltage is applied, the velocity of the ions in their mean free path becomes great enough to produce new ions by collision with atoms or molecules by separating the positive and negative parts of the atoms or molecules. These in turn produce new ions so the ionic density increases very rapidly. The positive ions travel to the negative conductor; the negative ions to the positive conductor. It seems, however, that the electron plays the principal part in impact ionization.

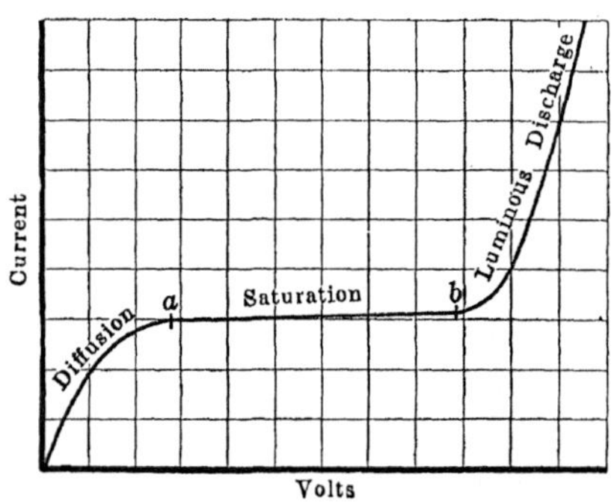

Fig. 163.—Variation of current with voltage through ionized space between parallel plate electrodes.

It may be of interest to review in a brief way how the law of visual corona was derived, to show that it is a rational law, and to illustrate a practical application of the electron theory.

In 1910 an investigation was conducted in which the visual corona voltages of various sizes of polished parallel wires at various spacings were determined.[1] For any given conductor the voltage gradient at rupture, g_v, was found to be constant independent of the spacing (except at very small spacings), but increased very rapidly as the radius of the conductor was decreased. Thus air *apparently* had a greater strength for small conductors than large

[1] F. W. Peek, Jr., Law of Corona, A.I.E.E., June, 1911.

ones. This was formerly pointed out by Prof. Ryan. The curve of experimental data between g_v and radius r was found to be regular and continuous. A number of equations could readily be written which would fit these data. It was desired, however, to establish or build up a rational equation. The old idea of air films at the surface of the conductors was abandoned after tests with very light (aluminum) and very heavy (tungsten) metals showed that the density of the metal of the conductor had no influence on g_v, which should be the case if the difference were caused by air films, as the air film should vary with the density of the conductor.

A rational law of visual corona was deduced from the data, as follows:

Energy of some form, be this the $\sum \frac{mv^2}{2}$ of the energy of the moving ions, or whatever form it may, is necessary to rupture insulation. This is borne out by experiments with transients, which show that finite time is necessary to rupture insulation; that if this time be limited the voltage must be increased to accomplish the same results in the limited time, also heating results at rupture, etc. These imply definite finite energy.

The gradient or stress to rupture air in bulk in a uniform field, g_o, should be constant for a given air density or molecular spacing. In a non-uniform field, as that around a wire, the breakdown strength of air, g_o, is first reached at the conductor surface; at an infinitely small distance from the conductor surface the stress is still below the rupturing gradient. Hence, in order to store the necessary finite rupturing energy the gradient at the conductor surface must be increased to g_v, so that at a finite distance away the gradient is g_o. This means that a finite thickness of the insulation must be under a stress of at least g_o. The rupturing energy is in the zone between g_v and g_o. The thickness of the zone should be a function of the conductor radius. It was found that at

$$0.301\sqrt{r} \text{ cm.}$$

from the conductor surface the gradient at rupture is always constant and 30 kv./cm. (at standard air density). The relation between g_v and r may now be directly expressed by the simple law

$$g_v = g_o\left(1 + \frac{0.301}{\sqrt{r}}\right) = 30\left(1 + \frac{0.301}{\sqrt{r}}\right) \text{ kv./cm. (max.)}$$

For parallel planes or for air in bulk

$$r = \infty$$
$$g_v = g_o$$

Thus the strength of air is constant and equal to 30 kv./cm., but in non-uniform fields is apparently stronger, as explained above. If the air is made less dense, that is the molecular spacing is changed, the strength of air in a uniform field should decrease directly with the air density.

$$g'_o = \delta g_o$$

However, for non-uniform fields, the energy distance should also change thus

$$0.301\sqrt{r}\phi(\delta)$$

The complete equation including air density factor was found to take the simple rational form

$$g_v = g_o\delta\left(1 + \frac{0.301}{\sqrt{\delta r}}\right)$$

This equation holds for values of δ as low as 0.02. As the air density is decreased, a minimum g_v is reached in the order of $\delta = 0.002$. g_v then increases very rapidly with decreasing δ. (See Fig. 164.) At very high vacuum the separation of the molecules may be of a fairly high order compared to the dimension of the tube. There is, thus, very little ionization by collision, and the apparent strength becomes high. At still higher vacuum, as used in the best X-ray tubes, the only source of ionization is from the conductors themselves. In the latest tubes the cathode is hot and is the only source of ions. The number can be controlled by the temperature of the cathode.[1]

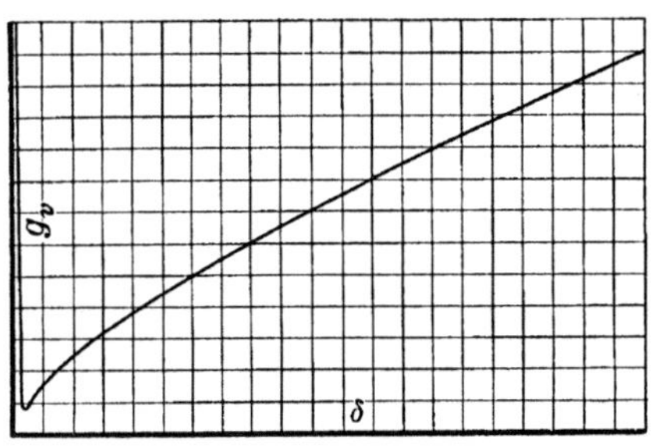

FIG. 164.—Variation of strength of air with density showing increase in strength at very low densities.

If the energy distance is limited to less than $0.301\sqrt{r}$, by placing the conductors close together, the apparent gradient should increase. This also is borne out by experiment. The visual tests

[1] Coolidge X-ray tube.

on various forms of electrodes, as spheres, wires, planes, etc., as well as loss measurements, all point to an energy storage distance and a constant strength of air of 30 kv./cm.

For continuously applied high *frequency* where the rate of energy or power is great, frequency may enter into the energy thus,

$$0.301\sqrt{\frac{r}{\delta}}\,\phi\,(f)$$

and spark-over take place at somewhat lower voltages. Where the time is limited, as by an impulse of steep wave front, a much higher voltage should be required to start arc-over. This is also borne out by experiment.

Thus far the form of this rupturing energy has not been considered as it was unnecessary, in order to develop a rational working equation.

When low potential is applied between two conductors any free ions are set in motion. As the potential and, therefore, the field intensity or gradient is increased, the velocity of the ions increases.[1] At the gradient of $g_o = 30$ kv./cm. ($\delta = 1$) the velocity of the ions becomes sufficiently great over the mean free path to form other ions by collision. This gradient is constant and is called dielectric strength of air. When ionic saturation is reached at any point, the air becomes conducting and glows, or there is corona or spark.

Applying this to parallel wires: when a gradient g_v is reached at the wire surface, any free ions are accelerated and produce other ions by collision with molecules, which are in turn accelerated. The ionic density is thus gradually increased by successive collision until at $0.301\sqrt{r}$ cm. from the wire surface, where $g_o = 30$, ionic saturation is reached, or corona starts. The distance $0.301\sqrt{r}$ cm. is, of course, many times greater than the mean free path of the ion, and many collisions must take place in this distance. Thus, for the wire, corona cannot form when the gradient of g_o is reached at the surface, as at any distance from the surface the gradient is less than g_o.

The gradient at the surface must therefore be increased to g_v so that the gradient a finite distance away from the surface ($0.301\sqrt{r}$ cm.) is g_o. That is to say, energy is necessary to start corona, as noted above. g_o the strength of air, should vary with

[1] See also explanation given in Theory of Corona, Bergen Davis, A.I.E.E., April, 1914.

δ; g_v, however, cannot vary directly with δ, because, with the greater mean free path of the ion at lower air densities, a greater "accelerating" or energy distance is necessary. In the equation, $a = 0.301\sqrt{r/\delta}$; that is, a increases with decreasing δ.

When the conductors are placed so close together that the free accelerating or energy storage distance is interfered with, the gradient g_v must be increased in order that ionic saturation may be reached in this limited distance.

Over a wide range, initial ionization of the air cannot affect the starting voltage of a steadily applied low-frequency e.m.f. since such ionization must necessarily be very small compared to the residual ionization after each cycle. If the ionization is very small when such a voltage is first applied an appreciable time is necessary before corona starts, but the starting voltage is not affected.

The initial ionization should, however, have a considerable effect upon transient voltages of short duration, steep wave-front voltages, etc. Thus, if for given electrodes, the time of application is less than that normally necessary to bring the initial ionization up to ionic saturation, a higher voltage should be required to cause corona or spark-over in the limited time. Experiments show, however, that even with impulse voltages the initial ionization may be varied over a great range without appreciable change in the impulse spark-over voltage of a given pair of electrodes in open air. The effect of initial ionization may be observed to a greater extent at low air densities, where the number of free ions may be made an appreciable per cent. of ionic saturation. When the electrodes are of such a nature as to require considerable brush discharge in the path of the spark before spark-over, the steep wave-front voltage required to cause discharge over a given gap is much higher than the required steady voltage. If spheres and needles, set to spark-over at the same 60-cycle voltage, are placed in parallel and impulse voltage of steep wave front applied, spark-over will take place across the spheres only.

CHAPTER IX

PRACTICAL CORONA CALCULATIONS FOR TRANSMISSION LINES

Summary of Various Factors Affecting Corona.—If potential is applied between the conductors of a transmission line and gradually increased, a point is reached when wattmeters placed in the circuit begin to read. The watt loss is low at first but increases with increasing voltages. At the point where the meters begin to read, a hissing noise is heard, and if it is quite dark, localized streamers and glow points can be noticed at spots on the conductors. These first streamers are caused by dirt and irregularities. The watt increase is still gradual. Slowly raising the voltage, the true visual critical corona point is finally reached. The effect is striking; corona suddenly jumps out, as it were, all along the line. The line immediately becomes very noisy, and the loss increases very rapidly with increasing voltage. The intensity of the light also increases with increasing voltage. If

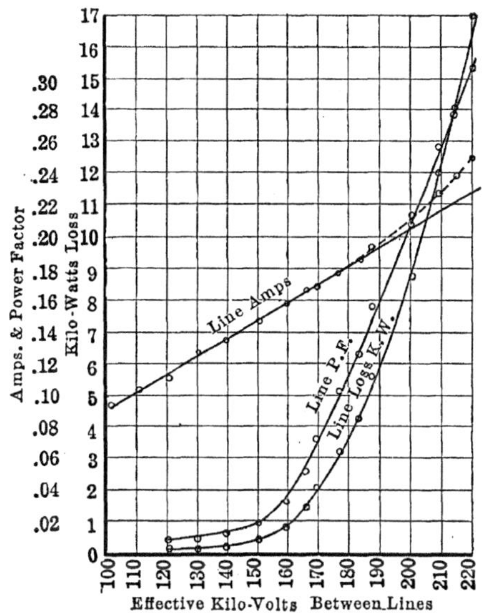

FIG. 166.—Characteristic corona loss, current, and power factor curves.
(Line A. Conductor length, 109,500 cm. (total). Spacing, 310 cm. Diameter, 1.18 cm. (3/0 cable). Temp., 12° C. Bar., 75.)

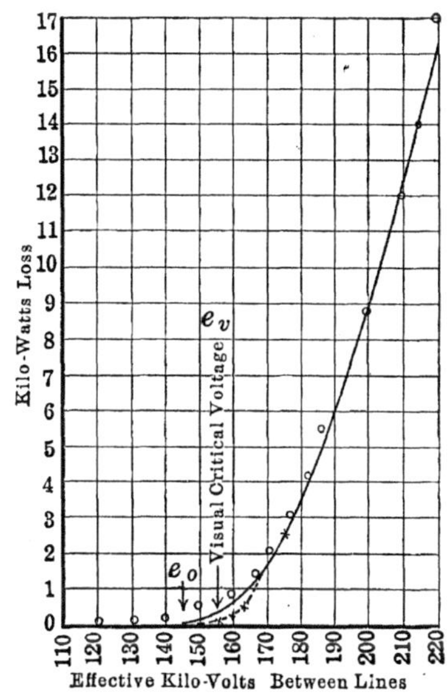

FIG. 167.—Loss near the critical voltage of new and old cables.
(Data Fig. 166. Line calculated. x, new cable. o, weathered cable).

one is close to the conductors an odor of ozone and nitrous oxide, the result of chemical action on the air, is noticed. There is, thus, heat, chemical action, light, and sound in the process of corona formation. These naturally all mean energy loss.

It becomes of great importance in the design of high-voltage transmission lines to know the various factors which affect the corona formation, and to have simple working formulæ for predetermining the corona characteristics, so that the corona loss will not be excessive.

Loss begins at some critical voltage, which depends upon the size and spacing of the line conductors, altitudes, etc., and increases very rapidly above this voltage. Figs. 166 and 167 show typical corona loss curves during fair weather.

An extensive investigation on an experimental transmission line (see Fig. 168) has shown that the corona loss in fair weather is expressed by the equation:

$$p = a(f + c)(e - e_o)^2$$

where p = loss in kilowatts per kilometer length of single-line conductor.

e = effective value of the voltage between the line conductors and neutral in kilovolts.[1]

f = frequency.

c = a constant = 25

and a is given by the equation

$$a = \frac{A}{\delta}\sqrt{r/s}$$

where r = radius of conductor in centimeters.

s = the distance between conductor and return conductor in centimeters.

δ = density of the air, referred to the density at 25 deg. C. and 76 cm. barometer as unity.

A = a constant = 241.

e_o = the *effective disruptive critical* voltage to neutral, and is given by the equation

$$e_o = m_o g_o \delta r \log_e s/r \text{ kv. to neutral}^1$$

where g_o is the *disruptive gradient of air in kilovolts* per centimeter at 25 deg. C. and 76 cm. barometer, and is constant for all

[1] Hence, in single-phase circuits, e is one-half the voltage between conductors. In three-phase circuits, e is $1/\sqrt{3}$ times the voltage between conductors.

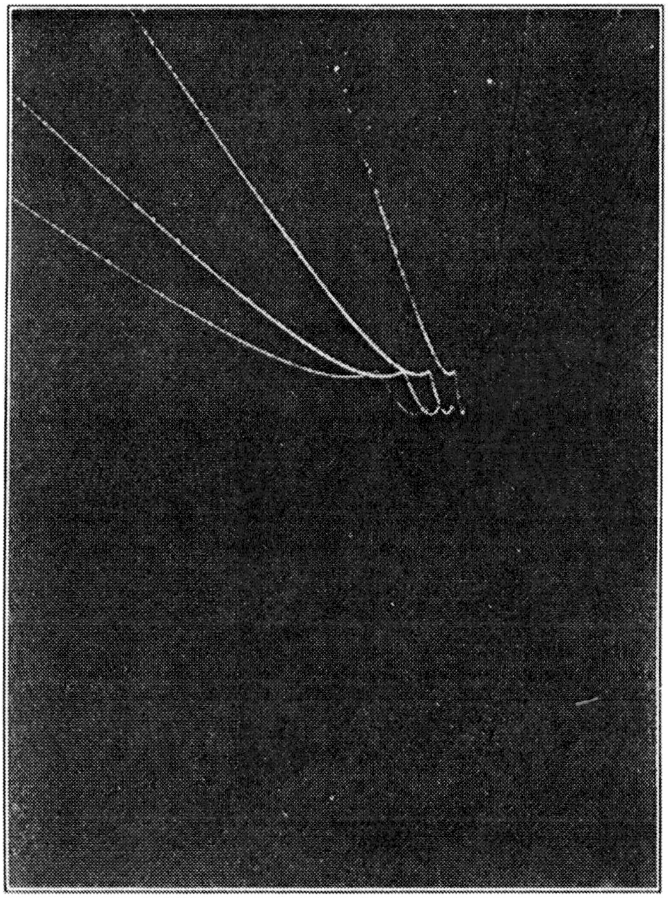

FIG. 165.—Corona at 230 kv. Line A. 3/0 cable. 310 cm. (122 in.) spacing.

(*Facing page* 200.)

FIG. 168.—Experimental transmission line.

practical transmission line sizes of conductor frequencies, etc. For very small conductors g_d is used. (See pages 136, 141, 142.)

g_o = 21.1 kv. per cm. (effective).

m_o is a constant depending upon the surface condition of the conductions, and is

m_o = 1 for perfectly smooth polished wire,

m_o = 0.98 to 0.93 for roughened or weathered wires, and decreases to

m_o = 0.87 to 0.83 for seven-strand cables (where the radius is taken as the outer radius of the cable).

Luminosity of the air surrounding the line conductors does not begin at the *disruptive critical voltage* e_o, but at a higher voltage e_v, the *visual critical voltage*. The *visual critical voltage*, e_v, is much higher for small conductors than the *disruptive critical voltage*, e_o; it is also higher for large conductors, but to a less extent. For very small conductors e_d replaces e_o. (See page 141.)

While theoretically no appreciable loss of power should occur below the visual voltage, e_v, some loss does occur, due to irregularities of the wire surface, dirt, etc., as indicated by brush discharges, and local corona streamers. (See page 143.)

As the loss between e_v and e_o depends upon dirt, roughened condition of the wire surface, etc., it is unstable and variable and changes as the surface of the conductors changes. For the larger sizes of stranded transmission conductors in practice the surface is generally such that the loss approximately follows the quadratic law, even between e_o and e_v. This is shown by the circles in Fig. 167, which are measured points on a weathered conductor. The crosses indicate how the points come on a new conductor. For a small conductor the difference between the calculated and measured losses on the section of the curve between e_o and e_v is still greater. Above e_v the curves coincide and follow the quadratic law. In practice it is rarely admissible to operate above the e_o voltage.[1] Cases are known in which conductors have deteriorated by the action of nitric acid formed by excessive brush discharge.

It is interesting to note that the loss below e_v actually follows the probability curve

$$p_1 = q\epsilon^{h-(e_0-e)^2}$$

but this need not be considered in practice.

[1] Operation at e_o voltage at high altitudes gives relatively greater margin and less loss than at sea level as there is greater difference between e_v and e_o.

The corona loss is:

(*a*) A loss proportional to the frequency f plus a small constant loss.

(*b*) Proportional to the square of the excess voltage above the disruptive critical voltage, e_o.

(*c*) Proportional to the conductor radius r and $\log_e s/r$. The critical voltage thus increases very rapidly with increasing r, spacing.

The *disruptive critical voltage*, e_o, is the voltage at which the disruptive voltage gradient of the air is reached at the conductor surface. Hence it is:

(*a*) Proportional to the conductor radius r and $\log_e s/r$. The critical voltage thus increases very rapidly with increasing r, and to a much less extent with increasing s.

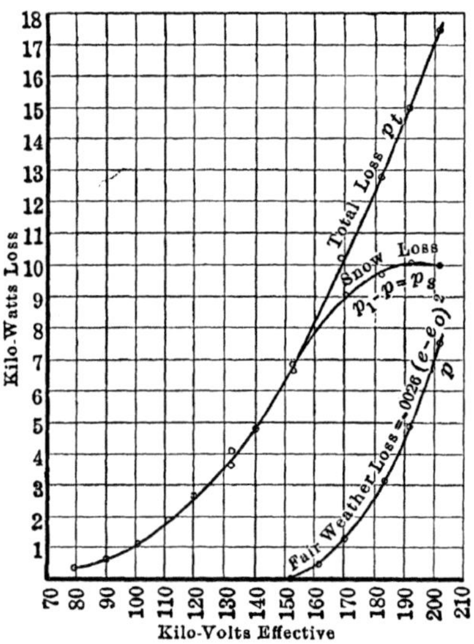

FIG. 169.—Comparison of fair weather and storm loss.

(*b*) Proportional to the air density or becomes very low at high altitudes.

(*c*) Dependent somewhat on the conditions of the conductor surface, as represented by m.

The effects of various atmospheric conditions and storms on the critical voltage and loss will now be considered.

(*a*) *Humidity* has no effect on the critical voltage.

(*b*) *Smoke* lowers the critical voltage and increases the loss.

(*c*) *Heavy winds* have no effect on the loss or critical voltage at ordinary commercial frequencies.

(*d*) *Fog* lowers the critical voltage and increases the loss.

(*e*) *Sleet* on the wires, or falling sleet, lowers the critical voltage and increases the loss. High voltages do not eliminate sleet formation.

(*f*) *Rain storms* lower the critical voltage and increase the loss.

(*g*) *Snow storms* lower the critical voltage and increase the loss.

(*h*) At high altitudes the loss is very much greater on a given conductor, at a given voltage, than it is at sea level. For a given voltage larger conductors must be used at high altitudes.

Fig. 169 shows a typical loss curve during storm and a corresponding fair weather curve.

Practical Corona Formulæ and Their Application.—The formulæ required for the determination of the corona characteristics of transmission lines are:

METRIC UNITS

Disruptive Critical Voltage

$$e_o = 21.1 m_o r\delta \log_e s/r \text{ kv. to neutral} \qquad (35)$$

Power Loss

$$p = \frac{244}{\delta}(f + 25)\sqrt{r/s}(e - e_o)^2 10^{-5} \text{ kw. per km. of single conductor} \qquad (34)$$

Where the conductors are small use formulæ (34*a*), (35*a*), and (36), Chapter V.

Visual Critical Voltage

$$e_v = 21.1 m_v \delta r\left(1 + \frac{0.301}{\sqrt{\delta r}}\right) \log_e s/r \text{ kv. to neutral} \qquad (20)$$

$$\delta = \frac{3.92b}{273 + t}$$

where

e = effective kilovolts to neutral applied to the line[1]

δ = air density factor

= 1 at 25 deg. C., 76 cm. barometer

b = barometric pressure in centimeters

t = temperature in degrees Centigrade

r = radius of conductor, centimeters

s = spacing between conductor centers in centimeters

f = frequency cycles per second.

For approximating storm loss consider e_o = 0.8 of fair weather value in formula (2).

Irregularity Factors

m_o = 1 for polished wires

= 0.98 – 0.93 for roughened or weathered wires

= 0.87 – 0.83 for seven-strand cables

m_v = m_o for polished wire = 1

m_v = 0.72 for local corona all along cable

m_v = 0.82 for decided corona all along cable.

[1] e = volts between line times $1/\sqrt{3}$ for three-phase lines and volts between line times ½ for single-phase or two-phase lines.

ENGLISH UNITS

Disruptive Critical Voltage

$$e_o = 123 m_o r \delta \log_{10} s/r \text{ kv. to neutral} \qquad (35')$$

Power Loss

$$p = \frac{390}{\delta}(f + 25)\sqrt{r/s}(e - e_o)^2 10^{-5} \text{ kw. per mile of single conductor} \qquad (34')$$

Where the conductors are small use formulæ (34*a*), (35*a*) and 36, Chapter V.

Visual Critical Voltage

$$e_v = 123 m_v \delta r\left(1 + \frac{0.189}{\sqrt{\delta r}}\right) \log_{10} s/r \text{ kv. to neutral} \qquad (20')$$

$$\delta = \frac{17.9b}{459 + t}$$

where

e = effective kilovolts to neutral applied to the line[1]
δ = air density factor
= 1 at 77 deg. F. and 29.9 in. barometer
b = barometric pressure in inches
t = temperature in degrees Fahrenheit
r = radius of conductor, inches
s = distance between conductor centers in inches
f = frequency, cycles per second.

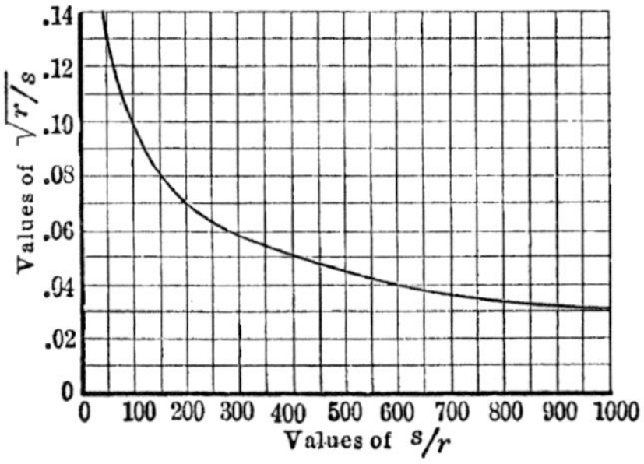

FIG. 170.

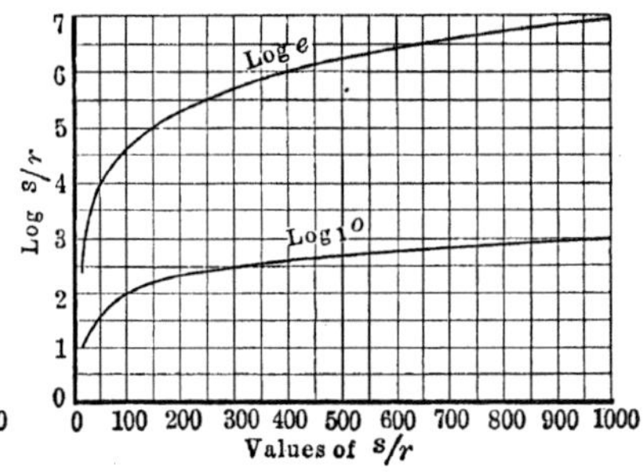

FIG. 171.

[1] e = volts between line times $1/\sqrt{3}$ for three-phase lines and volts between line times ½ for single-phase or two-phase lines.

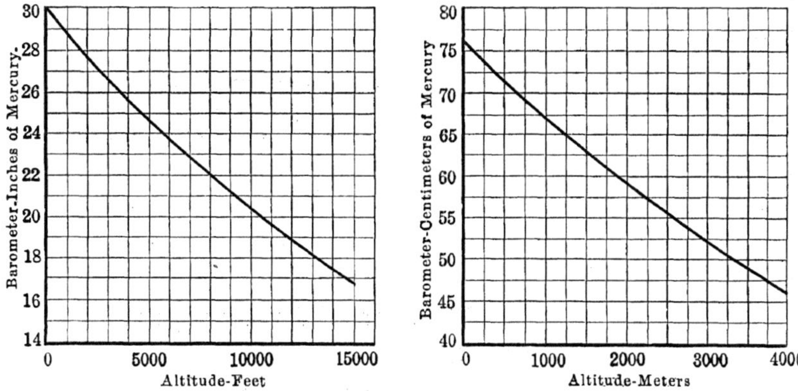

FIG. 172.—Approximate barometric readings at different altitudes.

In order to illustrate the use of the formulæ, the following practical example is taken:

EXAMPLE

Given: three-phase line	60 cycles
Length	100 miles
Spacing	120 in.
Conductor	1/0 cable—diam. 0.374 in.
Max. temperature	100 deg. F.
Elevation	1000 ft.
Barometer (from Fig. 172)	28.85 in.

Then $$s/r = \frac{120}{0.187} = 642$$

$$\log_{10} s/r = 2.81 \text{ (from Fig. 171 or tables)}$$

$$\sqrt{r/s} = 0.0394 \text{ (from Fig. 170 or calculated)}$$

$$\delta = \frac{17.9b}{459 + t} = \frac{17.9 \times 28.85}{459+100} = 0.925$$

$$m_o = 0.87$$

$$e_o = 123 m_o r \delta \log_{10} s/r = 123 \times 0.87 \times 0.187 \times 0.925 \times 2.81 \quad (35')$$

$$= 52.0 \text{ kv. to neutral}$$

$$p = 390(f + 25)\sqrt{r/s}(e - e_o)^2 10^{-5} \quad (34')$$

$$= 0.014(e - 52.0)^2 \text{ kw. per mile of single conductor.}$$

Since there are three conductors:

$$p = 0.042(e - 52.0)^2 \text{ kw. per mile of three-phase line.}$$

The conductor (100 deg. F., 1000 ft. elevation) would glow at

$$e_v = 123 m_v \delta r \left(1 + \frac{0.189}{\sqrt{\delta r}}\right) \log_{10} s/r = 62.6 \text{ kv. to neutral} \quad (20')$$

or $62.6 \times 1.73 = 108.3$ kv. between lines.

More decidedly at

$$71.3 \times 1.73 = 123.4 \text{ kv. between lines.}$$

The visual corona cannot be observed except on a very dark cloudy night.

To show the effect of altitude, the characteristics of the same line are calculated for 10,000 ft. elevation and tabulated in Table LXXIX. At this altitude

$$p = 0.0599(e - 36.8)^2 \text{ kw. per mile of three-phase line}$$
$$e_o = 36.8 \text{ kv. to neutral.}$$

$$(m_v = 0.72)\ 81.5 \text{ kv. local.}$$

Visual corona between lines

$$(m_v = 0.82)\ 92.7 \text{ kv. decided.}$$

TABLE LXXIX.—CORONA LOSS AT DIFFERENT OPERATING VOLTAGES

Kv. bet. lines	Kv. to neutral e	Kw. per mile and total in 100 miles							
		1000 ft. elevation				10,000 ft. elevation			
		Fair weather		Storms		Fair weather		Storms	
		p per mile, 3 cond.	100 miles	p per mile, 3 cond.	100 miles	p per mile, 3 cond.	100 miles	p per mile, 3 cond.	100 miles
50	28.9	......	....	......	....	......	....	......	....
60	34.7	......	....	......	....	......	....	1.7	170
70	40.5	......	....	......	....	0.8	80	8.0	800
80	46.3	......	....	0.9	90	5.4	540	17.0	1,700
90	52.0	0.0	0	4.6	460	14.0	1,400	30.0	3,000
100	57.8	1.4	140	11.0	1,100	26.0	2,600	48.0	4,800
110	63.6	5.7	570	20.0	2,000	33.0	3,300	70.0	7,000
120	69.4	13.0	1,300	33.0	3,300	63.0	6,300	96.0	9,600
130	75.1	22.0	2,200	47.0	4,700	88.0	8,800	125.0	12,500
140	80.9	35.0	3,500	65.0	6,500	116.0	11,600	158.0	15,800
150	86.7	51.0	5,100	86.0	8,600	149.0	14,900	196.0	19,600

For approximating the storm loss consider

$$e_o = 0.80 \text{ per cent. of its value in fair weather.}$$

Then $p = 0.042(e - 0.8 \times 52.0)^2 = 0.042\ (e - 41.6)^2$

This will give an idea of the maximum storm loss; it assumes a storm over the whole line at the same time, a condition that is most unlikely to occur. The storm loss will also generally be less due to lower temperatures.

The loss on a transmission line will vary from day to day depending upon the temperature and weather conditions. The above losses are calculated for summer temperature. For winter the losses would be much lower.

Safe and Economical Voltages.—It will generally be found that it is safe and economical to operate a line up to, but not above, the fair weather e_o voltage (determined for average barometer and 77 deg. F.). This gives a loss during storms of about 4.75 kw. per mile in the problem considered, but it is not likely that the storm will extend over the whole line at one time. Storm during cold weather will cause much less loss than the above. It thus is generally more economical to pay for the storm loss during small parts of the year rather than to try to eliminate it by an excessive conductor.

Thus for the above line a desirable operating voltage is in kilovolts between lines (1000 ft. elevation):

$$e_o \times 1.73 = 54.2 \times 1.73 = 93.8$$

(10,000 ft. elevation)

$$e_o \times 1.73 = 38.4 \times 1.73 = 66.5$$

It is undesirable to operate the conductors above the glow point as there is likely to be considerable chemical action on the conductor surface.

Methods of Increasing Size of Conductors.—The corona starting voltage is increased by increasing the diameter of the conductor. Thus, with the same amount of copper a hollow tube would give a higher corona voltage.

A hemp center conductor is not reliable.

For the same resistance an aluminum conductor has about a 25 per cent. greater diameter than a copper conductor and thus, approximately, 25 per cent. higher corona voltage.

In order to still further increase the advantage of aluminum, a steel-core aluminum cable has been put into operation in California at 150 to 180 kv. The critical voltage may also be decreased by grouping together conductors of the same potential.

Conductors not Spaced in Equilateral Triangle.—In three phase problems, it has been assumed that the conductors are arranged in an equilateral triangle. When the conductors are not spaced in equilateral triangle but, as is often the case in practice, symmetrically in a plane, corona will start at a lower voltage on the center conductor, where the stress is greatest, than on the outside conductor.

The actual critical voltage for the center conductor will be approximately 4 per cent. lower, and for the two outer conductors 6 per cent. higher, than the value for the same s, in the equilateral triangle arrangement.

If a triangle is used where there is considerable difference between s_1, s_2 and s_3, an exact calculation of the stress should be made. Such a calculation is quite complicated. (See Example Case 12, page 234.)

Voltage Change along Line.—For long lines the voltage, and therefore the corona loss, will vary at different parts of the line. This may be allowed for in a long line by calculating the loss per mile at a number of points. If a curve is plotted with these points as ordinates and length of line as abscissa the average ordinate may be taken as the loss per mile. The area of the curve is the total loss. A line operating very near the corona voltage may have no loss at load, but when the load is taken off there may be considerable loss due to the rise in voltage. This may sometimes be advantageous in preventing a very large rise in voltage when the load is lost. Grounding one *conductor* will also cause considerable loss on a line operating near the corona voltage by increasing the stress on the air at the conductor surface.

Agreement of Calculated Losses and Measured Losses on Commercial Transmission Lines.—It is generally difficult to make an exact comparison, as in most cases where losses have been measured on practical lines all of the necessary data, such as temperature along line, barometric pressure, voltage rise at the end of line, etc., has not been recorded. A range of voltage extending considerably above e_v is necessary to determine the curve. It is not generally possible to place such high voltages on practical lines.

The several examples given are ones in which it has been possible to obtain sufficient data to make a comparison. In Fig. 173 the drawn curve is calculated from the formulæ for the temperature, barometric conditions, etc., under which measurements were made. The crosses represent the measured values. This agreement between measured and calculated losses is very interesting, as it is for a long three-phase line at very high altitude.[1] (Note that the greater part of the curve is below e_v.) In Fig. 174 are similar comparisons on a single-phase line at different frequencies. These measurements were made before corona was a factor in practical transmission. An exact agreement could not

[1] Some of the first measurements on practical lines were made by Scott and Mershon, at high altitudes in Colorado. See A. I. E. E., 1898.

be expected, as the wave shape was not known and was probably not a true sine wave. Another rather interesting comparison is made in a recent publication describing the system of the Au Sable Electric Co.[1] The line of No. 0 copper cable is 125 miles long. With 140 kv. at the generating end, the voltage at the far end is 165 kv. Thus the loss per mile is greater at the far end than at the generating end. The average calculated loss per mile is 15 kw. The article states that the actual average loss by preliminary rough measurements was 15 to 20 kw. per mile. The above examples are given to illustrate how well the formulæ may be

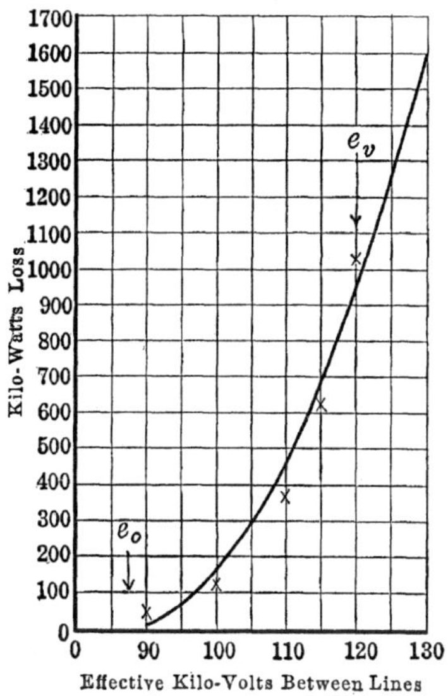

FIG. 173.

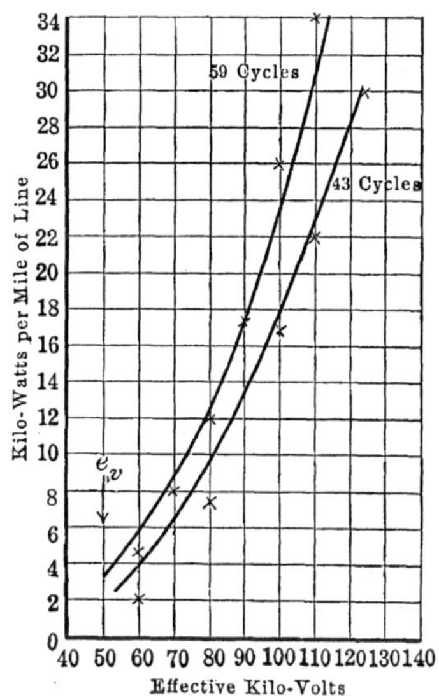

FIG. 174.

Comparison of calculated and measured losses.

FIG. 173.—Leadville. Three phase. (Test made by Faccioli.) 1/0 weathered cable. Spacing (flat) 124″ (314 cm.). Diameter, 0.95 cm. Length of line, 63.5 miles (one conductor). Temp., 11° C. Bar., 60 cm. Frequency, 60∼. Curve calculated. x measured.

FIG. 174.—Outdoor experimental line. (Tests by A. B. Hendricks, Pittsfield, 1906.) Wire conductor. Diameter, 0.064″ (radius 0.081 cm.). Spacing, 48″ (132 cm.). Length, about 800 ft. $\delta = 0.97$.

applied in the predetermination of the corona loss of commercial transmission lines.[2]

The Corona Limit of High-voltage Transmission, with Tables.—The question is often asked, what is the limiting distance imposed on high-voltage transmission by the corona limit of vol-

[1] N. E. L. A., June, 1912.

[2] Peek, F. W., Jr. Comparison of Calculated and Measured Corona Loss, A. I. E. E., Feb., 1915.

tage? In the first place it does not seem at present, except in very rare instances, that corona will be the limiting feature at all. The limiting feature will, in all probability, be an economic one; the energy concentrated naturally at any given point will be exceeded by the demand in the surrounding country before the transmission distance becomes so great that voltages above the corona limit are necessary. A quarter of a million volts may be used (except at high altitude) without an excessive conductor diameter or spacing (See Tables LXXX, LXXXI and LXXXII.)[1]

TABLE LXXX.—CORONA LIMIT OF VOLTAGE [2]

Kilovolts between Line Three-phase Sea Level and 25 deg. C.

Size B. & S. or cir. mils	Diameter, inches	Spacing, feet									
		3	4	5	6	8	10	12	14	16	20
4	0.230	...	56	58	60	62	64	66	68	69	71
3	0.261	...	62	65	67	70	72	74	76	77	80
2	0.290	...	...	71	73	76	79	81	83	85	87
1	0.330	...	...	79	81	85	88	91	93	95	97
0	0.374	...	...	...	90	95	98	102	104	108	109
00	0.420	...	...	...	98	104	108	111	114	117	121
000	0.470	...	...	...	...	114	118	121	124	127	132
0000	0.530	...	...	...	...	125	130	135	138	141	146
250,000	0.590	...	...	...	...	138	144	149	152	156	161
300,000	0.620	...	...	...	...	...	151	156	161	165	171
350,000	0.679	...	...	...	...	...	161	166	170	175	180
400,000	0.728	...	...	...	...	...	171	176	180	185	192
450,000	0.770	...	...	...	...	...	178	184	190	194	200
500,000	0.818	...	...	...	...	...	188	194	199	205	210
800,000	1.034	...	...	...	...	...	...	234	241	244	256
1,000,000	1.152	...	...	...	...	...	...	256	264	270	281

[1] These tables will give an approximate idea of the voltage limit imposed by corona. In general it will be found advisable to operate at or below the e_o voltage. Above e_o the storm loss becomes an important factor. Calculations should generally be made from formulæ for each special case.

To find the voltage at any altitude multiply the voltage found above by the δ corresponding to the altitude, as given in Table LXXXII.

[2] Kilovolts between lines corresponding to e_o at 25 deg. C. and 76 cm. barometer. Conductors arranged in an equilateral triangle.

TABLE LXXXI.—CORONA LIMIT OF VOLTAGE [1]

Kilovolts between Line Three-phase Sea Level and 25 deg. C.

WIRES

Size B. & S. or cm.	Diameter, inches	Spacing, feet									
		3	4	5	6	8	10	12	14	16	20
4	0.204	51	54	56	58	60	62	64	65	66	68
3	0.229	...	59	62	64	66	68	70	72	74	76
2	0.258	...	...	69	70	74	76	78	80	82	84
1	0.289	...	...	75	77	81	83	86	88	90	92
0	0.325	...	...	...	85	89	92	95	97	99	102
00	0.365	...	...	...	94	98	102	105	107	110	113
000	0.410	...	...	...	...	109	113	116	119	121	124
0000	0.460	...	...	...	...	120	125	128	131	134	138

To find the voltage at any altitude multiply the voltage found above by the δ corresponding to the altitude, as given in Table LXXXII.

For single phase or two phase find the three-phase volts above and multiply by 1.16.

TABLE LXXXII.—ALTITUDE CORRECTION FACTOR AT 25 deg. C.

Altitude, feet	δ	Altitude, feet	δ
0	1.00	5,000	0.82
500	0.98	6,000	0.79
1,000	0.96	7,000	0.77
1,500	0.94	8,000	0.74
2,000	0.92	9,000	0.71
2,500	0.91	10,000	0.68
3,000	0.89	12,000	0.63
4,000	0.86	14,000	0.58

Corona is generally thought of as affecting only very high voltage transmission lines and of little other practical importance. It may, however, exist at almost any voltage, as it depends not only upon voltage but also upon the configuration of the electrode and the spacing. It will exist in cables, on insulators, coils,

[1] Kilovolts between lines corresponding to e_o at 25 deg. C. and 76 cm. barometer.

switches, etc.; in short wherever air is present, and where damage may be done by mechanical bombardment of small streamers, and by electrochemical action (ozone and nitrous oxide), unless means are taken to prevent it. In most cases its prevention is fairly simple. This is further treated in Chapter X.

CHAPTER X

PRACTICAL CONSIDERATION IN THE DESIGN OF APPARATUS WHERE SOLID, LIQUID AND GASEOUS INSULATIONS ENTER IN COMBINATION

As important as the quality of the dielectric is the configuration of it and of the electrodes. It is also of importance in combining dielectrics of different permittivities to see that one does not weaken the other by causing unequal division of stress. It is possible to cause breakdown in apparatus by the addition of good insulation, dielectrically stronger than the original insulation.

Case 1. Breakdown Caused by the Addition of Stronger Insulation of Higher Permittivity.—As an example take two

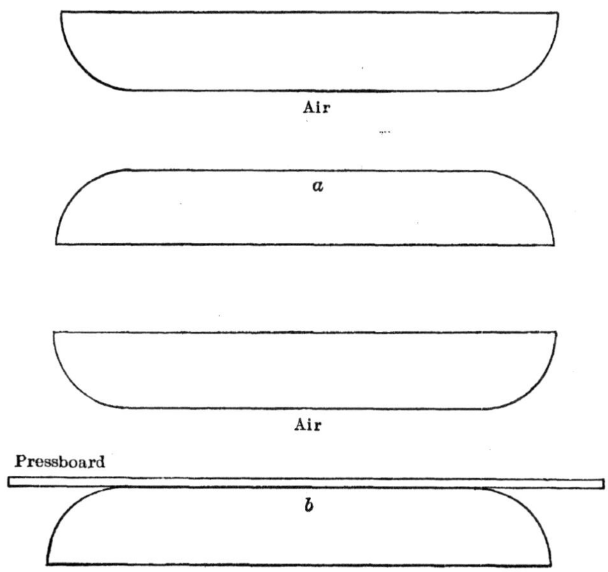

Fig. 175.—Causing break-down by the addition of insulation.

parallel plates rounded at the edges and placed 2 cm. apart in air, as in Fig. 175(*a*).

(1) Apply 60 kv. (max.) between the plates. The gradient then is $60/2 = 30$ kv./cm. max. (neglecting flux concentration at edges). As a gradient of 31 kv./cm. max. is required to rupture air there is no breakdown.

(2) Remove the voltage and insert between the plates a sheet of pressboard 0.2 cm. thick, as in Fig. 175(*b*). The constants of pressboard are

$$k = 4 \text{ rupturing gradient} = 175 \text{ kv./cm. max.}$$

$$l_p = 0.2 \text{ cm.}$$

For air

$$k = 1 \text{ rupturing gradient} = 31 \text{ kv./cm.}$$

Apply now the same voltage as before. The addition of the pressboard of higher permittivity has increased the capacity of the combination and, therefore, the total flux and the flux density in the air. The gradient has also been increased. The combination may be considered as two condensers in series.

Where c_1 = capacity of air condenser, c_2 = capacity of pressboard condenser; e_1 = voltage across the air condenser, e_2 = voltage across the pressboard condenser; the total flux is

$$\psi = c_1e_1 = c_2e_2$$

$$\psi = \frac{k_1A}{l_1} e_1 = \frac{k_2A}{l_2} e_2$$

therefore, $\frac{k_1}{l_1} e_1 = \frac{k_2}{l_2} e_2$

$$e = e_2 + e_1$$

$$\frac{k_1}{l_1} e_1 = \frac{k_2}{l_2}(e - e_1)$$

$$e_1\left(\frac{k_1}{l_1} + \frac{k_2}{l_2}\right) = \frac{k_2e}{l_2}$$

$$e_1 = \frac{\frac{k_2e}{l_2}}{k_1/l_1 + k_2/l_2} = \frac{l_1k_2e}{k_1l_2 + k_2l_1} = 58.4$$

$$e_2 = 60 - 58.4 = 1.6$$

$$g_1 = \frac{e_1}{l_1} = \frac{k_2e}{k_1l_2 + k_2l_1}$$

Then $$g_1 = \frac{4 \times 60}{1 \times 0.2 + 4 \times 1.8} = 32.5 \text{ kv./cm. max.}$$

The air breaks down as g_1 is higher than the critical gradient, causing breakdown of air. As the broken down air is conducting, most of the applied voltage is placed on the pressboard. Thus, after the air ruptures, the gradient on the pressboard is:

$$g'_2 = \frac{60}{0.2} = 300 \text{ kv./cm. max.}$$

This is much greater than the rupturing gradient of pressboard and causes it to break down. Therefore, the 2.0-cm. space, which is safe with air alone, is broken down by the addition of stronger insulating material of higher permittivity. The stresses on this combination could have been calculated directly from (16a), Chapter II.

Another and convenient way of looking at this is as follows:

$$\text{Flux} = \frac{\text{volts}}{\text{"flux resistance" or "elastance"}}$$

S is termed the elastance and is the reciprocal of permittance. The reciprocal of the permittivity is termed the elastivity. For the given electrode arrangement, S is proportional to the elastivity and the length, as long as lines of force are practically straight lines. Let the relative elastivities be

$$\begin{aligned} \text{Air} &= 1.0 \\ \text{Pressboard} &= 0.25 \end{aligned}$$

Let the absolute elastivity of the air be σ_a, and introduce for the sake of abbreviation $a = \sigma_a/A$, where A is the area.

Then the elastances for the different circuits in the test are:

(*a*) Air (2 cm.)	$S_1 = a \times 2 \times 1$	$= 2a$
(*b*) Pressboard (0.2 cm.)	$S_p = a \times 0.2 \times 0.25$	$= 0.05a$
(*c*) Air (1.8 cm.)	$S_a = a \times 1.8 \times 1$	$= 1.8a$
(*d*) 1.8 cm. air + 0.2 cm. pressboard	$S = S_p + S_a$	$= 1.85a$

Thus when there is only air in the gap

$$\psi_1 = \frac{60}{2a} = \frac{30}{a}$$

When 0.2 cm. of air is removed and the same thickness of pressboard added, the total elastance is less and the flux increases or

$$\psi_2 = \frac{60}{1.85a} = \frac{32.5}{a}$$

The "drop" across the air is

$$\psi_2 \times S_a = \frac{32.5}{a} \times 1.85a = 58.4 \text{ kv.}$$

The "drop" across the pressboard is

$$\psi_2 \times S_p = \frac{32.5}{a} \times 0.05a = 1.6 \text{ kv.}$$

58.4 kv. is sufficient to cause 1.8 cm. of air to break down. When the air breaks down the full 60 kv. appears across the pressboard which, in turn, breaks down.

The case discussed above is an exaggerated example of conditions often met in practice. In many power stations little bluish needle-like discharges, called "static," may be noticed around generator coils, bushings, etc. This "static" is simply overstressed or broken down air, but unlike Case 1, the solid dielectric is sufficiently thick so that very little extra stress is put upon it by the broken down air. Damage may be caused in the course of time, however, by local heating, chemical bombardment, etc.

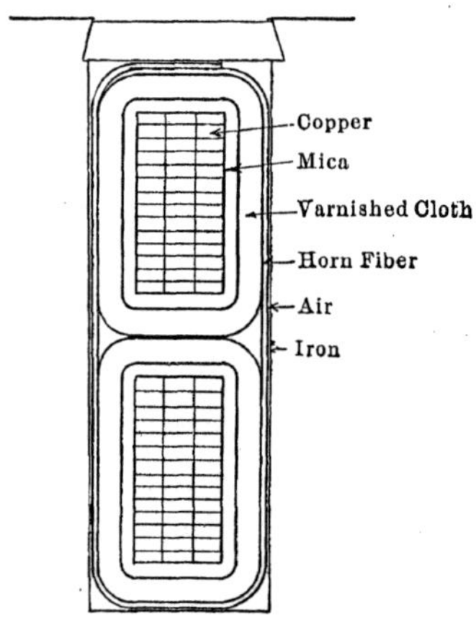

FIG. 176.—Corona on generator coils.

Case 2. Static or Corona on Generator Coils.—Consider the terminal coil of a 13,200-volt generator, insulated by 0.25 cm. of built-up mica ($k = 4$), and 0.45 cm. of varnished cambric ($k = 5$). In the slot is an armor of 0.1-cm. horn fiber ($k = 2.5$). In series with these is more or less air; assume 0.05 cm. for the purpose of this calculation. A section of the coil assembled in the machine is shown in Fig. 176. The stress on the air may be approximately found by assuming the conductors as one flat plate of a condenser, and the frame as the other.

Then:

Kilovolts between lines	13.20	eff.
Kilovolts to neutral	7.63	eff.
Kilovolts to neutral	10.7	max.

$$g_{air} = \frac{e}{k_4\left(\frac{x_1}{k_1} + \frac{x_2}{k_2} + \frac{x_3}{k_3} + \frac{x_4}{k_4}\right)} \quad (16a)$$

$$= \frac{10.7}{1\left(\frac{0.25}{4} + \frac{0.45}{5} + \frac{0.1}{2.5} + \frac{0.05}{1}\right)}$$

$$= \frac{10.7}{0.242} = 44 \text{ kv./cm. max.}$$

Since the disruptive strength of air is 31 kv./cm., it will break down, forming corona. Experience has shown that in time the corona eats away the insulation by mechanical bombardment, local heating, and chemical action, and ultimately a short circuit results.

Assume that the machine is operated at 8000 volts. The air is then stressed to 26.7 kv./cm. and corona does not form. However, should one phase become grounded, the voltage of the other two above ground would become 8.0 kilovolts instead of 4.62 kilovolts, the gradient on the air would rise to 46.3 kv./cm., and corona would result. Assume now that there are no grounds, but that the machine, which shows no corona at 8.0 kilovolts at sea level, is shipped to Denver. The altitude of Denver is approximately 5000 ft. The corresponding barometer is 24.5 cm. (Fig. 172), and hence the relative air density at 25 deg. C. is 0.82. At this density the disruptive strength of air is $0.82 \times 30 = 25.5$ kv./cm. The air around the coils near the terminals having a gradient of 26.7 kv./cm. would glow.

The best method of preventing this corona on machine coils is to tightly cover the surface of the coil with a conductor, as tinfoil, and to connect the foil to the iron frame of the machine; this, in effect, short circuits the air space. Naturally, the foil should be slit in such a way as to prevent it from becoming a short-circuited turn by transformer action.

Case 3. Overstressed Air in Entrance Bushings.—Assume that a ¾-in. conductor, supplying power at 33 kv., enters a building through a 3-½-in. porcelain bushing having a 1-in. hole (see diagram Fig. 177). The voltage between the rod and the ground ring is 19 kv. The stress on the air at the surface of the rod is:

$$r_1 = 0.375 \text{ in.} \qquad k_1 = 1$$
$$r_2 = 0.5 \text{ in.} \qquad k_2 = 4$$
$$R = 1.75 \text{ in.}$$

$$g_{air} = \frac{e}{xk_x\left(\dfrac{\log_e \dfrac{r_2}{r_1}}{k_1} + \dfrac{\log_e \dfrac{R}{r^2}}{k_2}\right)} \qquad (17)$$

$$= \frac{19 \times 1.41}{(0.375 \times 1)\left(\dfrac{\log_e 1.33}{1} + \dfrac{\log_e 3.5}{4}\right)}$$

$$= 120 \text{ kv./in.}$$

or $\quad 47.5$ kv./cm. max.

To cause corona on a rod of this size a gradient

$$g_v = 31\left(1 + \frac{0.3}{\sqrt{r}}\right)$$
$$= 31\left(1 + \frac{0.3}{\sqrt{0.375 \times 2.54}}\right)$$
$$= 40.5 \text{ kv./cm. max.}$$

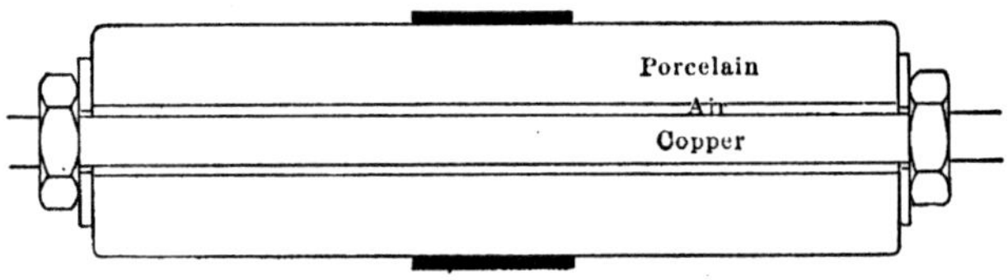

FIG. 177.—Corona in bushing.

is necessary. Hence, in the case considered, there will be corona, and chemical action on the rod which will become coated with a green surface of copper nitrate. The obvious cure for this is to coat the inside of the porcelain shell with a conductor and connect it to the rod.

Corona or "static" is often noticed where insulated cables come through a wall or bushing. For instance, three rubber covered cables may come through three bushings in a wall. If the voltage is high enough a glow will appear around the rubber in the air space inside the bushing. Ozone attacks rubber very rapidly. Such cables may soon be broken down by this simple cause. Such breakdowns are often ascribed to "high frequency." The remedy is to "short circuit the air space." In doing this by metal tubes slit lengthwise, to prevent eddy loss in the metal, care must be taken to bell the ends of the tube, otherwise the air will be stressed where the metal tube ends. The "belled" part may be filled with solid insulation.

Case 4. Graded Cable.—Assume that three insulations are available, all of exactly the same dielectric strength, but of permittivities as follows:

$$\text{Insulation } A,\ k = 5.4$$
$$\text{Insulation } B,\ k = 3.6$$
$$\text{Insulation } C,\ k = 2.0$$
$$\text{Rupturing gradient, } g = 100 \text{ kv./cm.}$$

Assume that it is desired to insulate a 1.0-cm. wire using 1.75 cm. of insulation with an outside lead cover. The best way of applying the insulation is so that each part is stressed in proportion to its respective strength. This ideal cable is impossible, but the more nearly this condition is realized the higher the voltage that may be applied to the cable without rupture.

(*a*) Using insulation *A* alone the breakdown voltage is

$$e = gr \log_e R/r = 75 \text{ kv.}$$

This is the same for either *B* or *C* alone.

(*b*) Using insulation *A* next to the wire, then *B*, then *C*, with

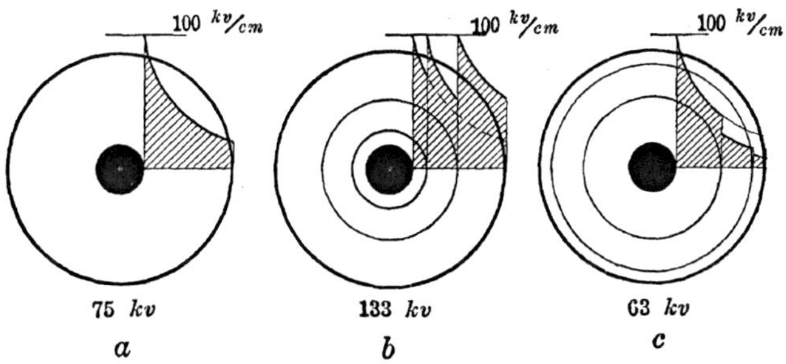

FIG. 178.—Graded cable.
(*a*) Not graded. (*b*) Insulations properly arranged. (*c*) Insulations improperly arranged. Noted break-down voltage in each case.

thicknesses 0.25 cm., 0.6 cm., and 0.9 cm. respectively, as in Fig. 178*b*, the rupturing voltage is

$$e = 133 \text{ kv.}$$

(*c*) Using the insulations in the reverse order as in Fig. 178*c* the rupturing voltage is

$$e = 63 \text{ kv.}$$

Note that the area in Fig. 178*a*, *b*, and *c* represents the voltage, and therefore the rupturing voltage if the maximum *g* is the rupturing gradient. This example is given to show how important it is to properly arrange insulations. In general the insulation of the highest permittivity should be placed where the field is densest. This applies not only to cables but all electrical apparatus. In spite of the fact that all of the above insulations had the same dielectric strength, and the same total thickness, the rupturing voltages with the different arrangement were 133 kv., 75 kv., and 63 kv. respectively. (Use equation (17).)

Case 5. Bushing.—Other cases where the principle of putting insulation of high permittivity at points of dense field is shown are illustrated in Figs. 18 and 179. The solid insulation of the lead in Fig. 18, because its contour follows the lines of force, does not increase the stress on the air near it. It is, for that reason, much better than an insulator which has the insulation arranged in such a way that the stress on the air is increased in the denser part of the field. However, by inserting the high permittivity insulation in the dense fields at the rod, and cutting it away in the middle, a better arrangement is obtained (Fig. 179). In the zone between flux lines a and b, there is now the insulation of high permittivity, and air of low permittivity, in series in the same way as in the graded cable, and with the same effect. In tests the insulation of the type of Fig. 179 arced over at 18 per cent. higher voltage than that of Fig. 18 (of same over-all dimensions), though its air path was 6.5 per cent. shorter.[1]

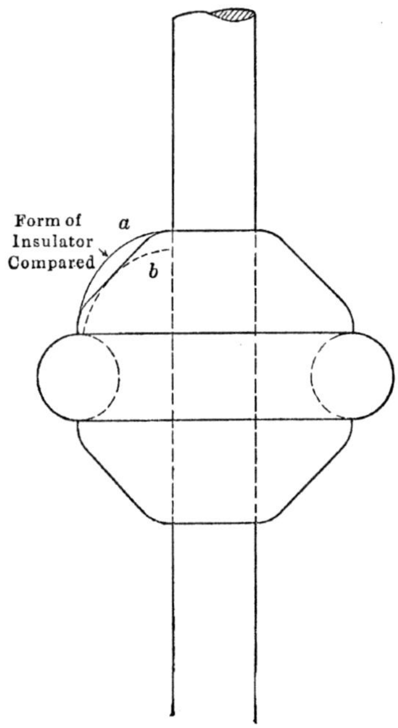

FIG. 179.—Bushing. Rod and Torus.[2]

In practice it is generally necessary to add corrugations to increase the "leakage path" on account of dirt settling on the surface, etc. The "ideal" design is not always the best in practice.

Case 6. Transformer Leads or Bushings.—One of the most common bushings is the oil filled type. In the design of such a bushing two general problems present themselves: The internal stress on the oil, which determines the puncture voltage; the external stress on the air, which determines the arc-over voltage.

In the design of a bushing, however, the whole dielectric circuit must be considered at the same time.

If the surface of the shell follows a line of force, the internal field does not cut the shell and cause flux concentration at points on the shell; the voltage per unit length of surface,

[1] Fortescue, Paper, A.I.E.E., March, 1913.
Weed, Discussion, A.I.E.E., March, 1913.

[2] Figs. 18 and 179 cannot represent practical leads as the rod and torus field is changed by the plane of the transformer case.

however, is not constant with this condition unless the lines of force are approximately straight parallel lines A bushing, when parallel planes are approximated, theoretically, need not be over 2 in. high for a 100-kv. arc-over. This imposes a large diameter compared to length, and large well-rounded metal caps, etc. Such a bushing would arc over with slightly dirty surfaces, moisture on surfaces, etc., at very low voltage. A practical lead must generally be fairly long and corrugated (See Table LXXVIII, page 190).

Where the surface follows a curved line of force, the internal field still does not cut the surface and cause the so-called "leakage" by local flux concentration, but the gradient is not constant along this line of force. Local breakdowns precede spark-over.

If the lead does not follow a line of force, the lines from the outside pass through the shell to the inside. In this case the shell should be so shaped that the stress is divided parallel to the surface and also perpendicular to the surface.

In an improperly designed bushing of this sort, breakdown might occur at places along the surface, and at other points out from the surface. For instance, as an extreme case, it might be imagined that the surface of a bushing followed an equipotential surface. There would then be no stress in the direction of the surface, but breakdown would occur by the stress perpendicular to the surfaces, as corona. Conversely a condition might obtain which would cause greatest stress along the surface as when a line of force is followed. As the surface, due to dirt, etc., is generally weaker than the air, it is in most cases better not to have maximum stress along it. (See Table LXXVIII.) In any case the stress should be uniform measured in either direction. Where the shells follow a line of force the field is more readily approximated by experiment, or by calculation, than where the lines of form cut the shell, when flux refraction, etc., must be considered. The direction of the flux may be controlled by the arrangement of metal parts. The bushing problem is a space problem.

In practice it is generally necessary to add petticoats. These should be placed so as to produce the minimum disturbance in the field with minimum stress along their surfaces.

Another type of lead is the condenser lead, built with the *object* of stressing all of the solid insulating material approximately equally. It consists of a number of cylindrical condensers of equal thickness, but of unequal lengths, arranged in such a way as to make the several capacities equal (Fig. 180). If this were exactly the case, the voltages across the equal thicknesses of insulation and equal distances along with the surface would be equal. This condition is possible but not generally reached on account of the capacity of the condensers to ground, to the central rod, and to each other (Fig. 180); to secure equal division of voltage here (Fig. 180) it would be necessary to connect the condenser plates to proper sources of potential. The condition may be approached, however, to such a degree that a good practical lead is the result when the insulating has been carefully done. A smaller diameter is obtained, but a greater length is re-

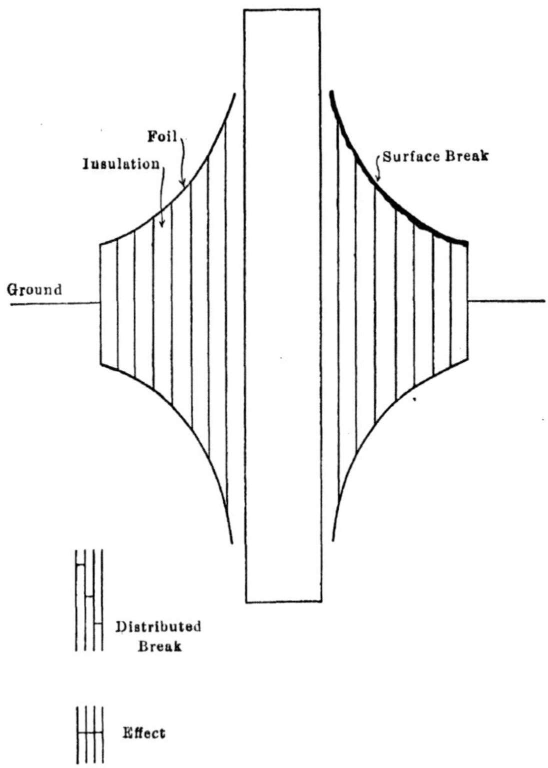

FIG. 180.—Condenser bushing.

quired due to arcing distance in the air (*i.e.*, to avoid arcs of the nature of the heavy line, Fig. 180). The present practical form is a long thin lead. (One disadvantage claimed is shown in the small sketch accompanying Fig. 180.) Little flaws in various parts of the lead are lined up by the metal parts and put directly in series. The separation of the insulation by metal is, from another standpoint, an advantage. A progressing corona streamer is stopped on reaching the metal surface at any layer. Concentration of flux at the edges of the metal cylinder must also be taken care of. The condenser lead is often arranged with a metal hat for flux control.

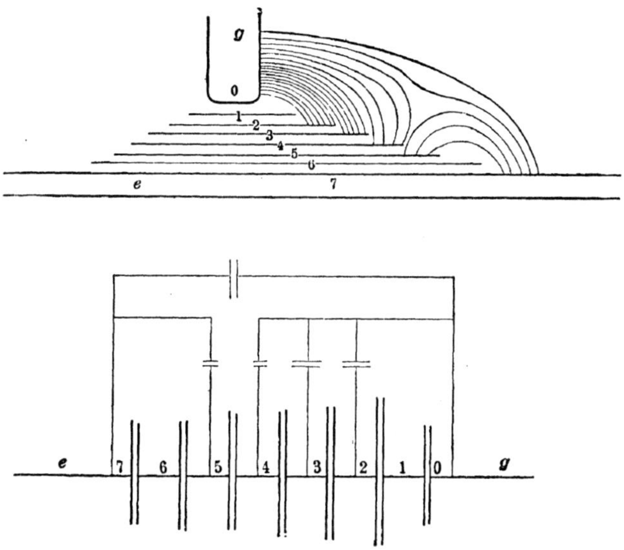

FIG. 181.—Diagramatic representation of flux, and capacities in condenser bushing.

Case 7. Dielectric Field Control by Metal Guard Rings, Shields, Etc.—It is sometimes practical to accomplish more with metal than by added insulation. In a case where the field is not uniform, but very much more dense at one point than at another, the flux may be made more uniform by relieving the dense portion and distributing over the less dense portion by a proper arrangement of metal parts connected to a source of potential of the proper value. This is not always practical, as the necessary complicated potential connections often weaken the apparatus and make it much more liable to breakdown.

As a simple example: Fig. 182 represents two spheres of unequal size in air, one at potential e the other at potential $-e$, or a voltage $2e$ between them. All the flux from A ends on B. The flux density at B is then much greater than at A and the air around B is very much more stressed than the air at the surface of A. The equipotential surface C may be covered with thin metal and no change takes place in the flux at A or B. If, however, C is connected by a wire to B the flux around B disappears and there is no stress on the dielectric at the surface of *B*. The total flux increases because of the greater capacity between *A* and *C*. The stress at *A*, due to increased flux density, increases, but it is still much less than the stress formerly at *B*, and a greater potential is required for spark-over. In other words, the insulation is more uniformly stressed and therefore working at greater efficiency.

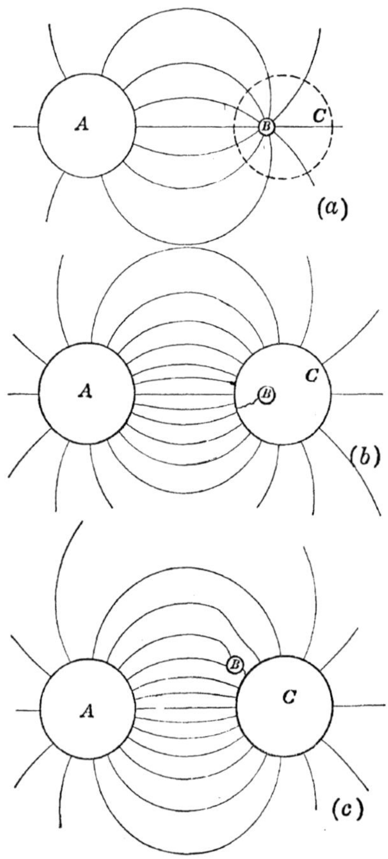

FIG. 182.—Simple illustration of flux control.

If instead of surrounding *B* or completely shielding it the sphere *C* be placed as in (*c*) and connected to *B* by a wire the stress is relieved at *B* and increased at *A*. The distribution at *B* is again more uniform.

An actual example where this principle was made use of in an emergency case several years ago by the author is illustrated in Figs. 183 and 184. In making some experiments 200 kv. were carried through the roof of a shed by porcelain bushings. During a heavy wind storm the roof was blown off and one bushing cracked as indicated by the jagged line. When the roof was replaced and an attempt made to put the bushing again into operation it was found that bad arcing took place to the damp wood at 130 kv. As no extra bushings were immediately obtainable, and it was necessary to finish the experiments, the expedient of field control was made use of. By hanging a metal torus made of a coil of wire on the rod, the work of about fifteen minutes, the bushing was made operative up to 200 kv. This was not as good as a new bushing, not the best sort of bushing, but it was a means of making a defective bushing operative, and prevented a shutdown of a month or more.

The effect of this shield is shown diagrammatically in Fig. 184.

By moving the ring up and down the rod, a point of minimum flux density on the cracked surface of the porcelain is found. It better distributed the flux and reduced the maximum flux density below the rupturing value.

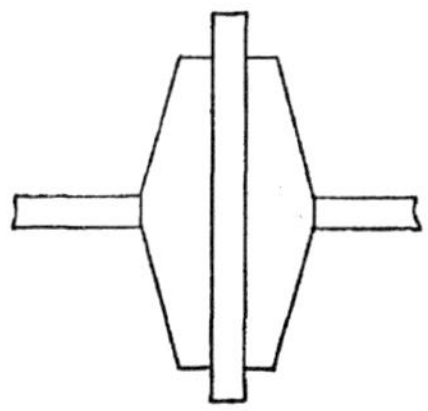

FIG. 183.—Entrance bushing.

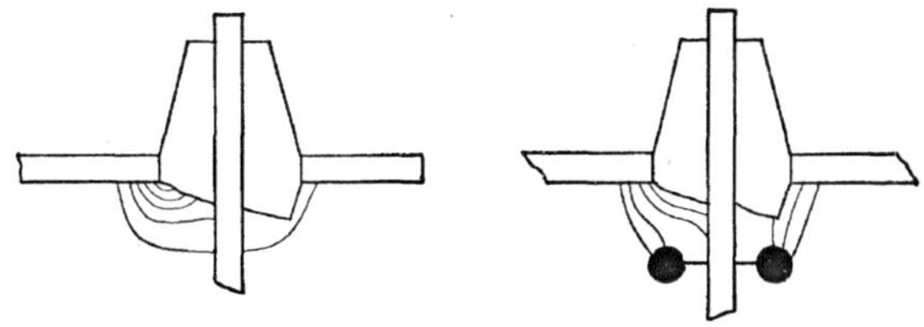

FIG. 184.—Making bushing shown in Fig. 183 operative by flux control after lower part had broken off.

Case 8. High Frequency.—Apparatus must be designed to meet not only normal but also, to a reasonable extent, abnormal conditions. Due to surges, lightning, arcing grounds, switching, etc., high frequency voltages travel over the line, say to a transformer. The voltage may not be increased at the transformer terminals, as the "high frequency" may exist only as a slight ripple on normal voltage wave. The lightning arrester therefore does not discharge, and a needle gap across the transformer terminals shows no voltage rise. A needle gap across a small section of the transformer may indicate a voltage several times normal line voltage. The points of greatest potential difference will depend upon the frequency and the transformer constants. The fact that "high frequency" may enter an apparently high inductance and build up high local potentials is because the inductance also contains capacity. It is not possible to give a theoretical treatment of this here. It is simply mentioned to show that it is sometimes dangerous to thin insulation at one part and add it at another in order to get perfect flux distribution under normal conditions, because under abnormal conditions great potential differences may exist across the weak insulation. It also illus-

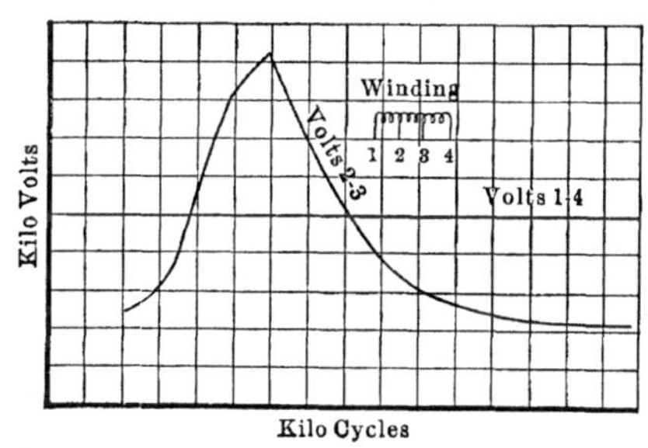

FIG. 185.—Effect of "high frequency" on voltage distribution across a coil of wire.

trates how "high frequency" generally is dangerous by building up high local potential differences. Where a large number of metal parts are used to distribute the flux under normal condition, the effect of the well-known multi-gap lightning arrester may come in and cause break-down at high frequencies.

It is, for somewhat similar reasons, not always best to follow ideal designs in line insulators, leads, etc. The ideal surface may be such as to make the surface time lag low or the rain arc-over low.

Case 9. Dielectric Field.—Draw the dielectric lines of force and equipotential surfaces between two parallel cylinders so that $\frac{1}{12}$ of the flux is included between any two adjacent lines of force, and $\frac{1}{20}$ of the voltage is between any two adjacent equipotential circles.

Let $S = 10$ cm. between conductor centers
$r = 1$ cm. = conductor radius.

From Chapter II, page 21,

$$z = \frac{S - \sqrt{S^2 - 4r^2}}{2} = \frac{10 - \sqrt{100 - 4}}{2} = 0.10$$

The distance between focal points of the lines of force is

$$S' = S - 2z = 10 - 0.20 = 9.80$$

It is desired to include one-twelfth of the total flux between lines of force. Draw radial lines from the flux centers $\frac{1}{12} \times 180 = 30$ deg. apart. (See Chapter II, pages 14 and 19.) The point of intersection of a radial line with $N. N.$ (Fig. 186*a*) is a point on the line of force. The line of force is hence a circle with center on $N. N.$, and passing through the point of intersection of the radial line $N.N.$, A'_1, and A'_2. The lines of force are therefore determined. The centers, etc., might have been calculated from equation (8), page 19. The line of force is also determined graphically by drawing the diagonal line through the intersection of radial lines, as shown in Fig. 187.

The equipotential surfaces in this case may be found graphically in a similar way by drawing diagonals through the intersections of the circles of the component fields. Other resultant fields may be drawn from corresponding plane diagrams if the component fields are given for the same strength, or same potential differences between the equipotential surfaces. Any number of fields may be so combined, two at a time. It is often possible to approximate for practical purposes the field of a complicated structure by so combining the simple fields of the component electrodes.

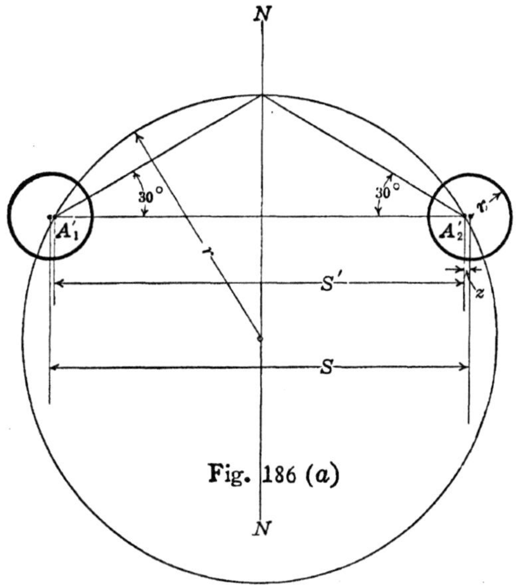

Fig. 186 (a)

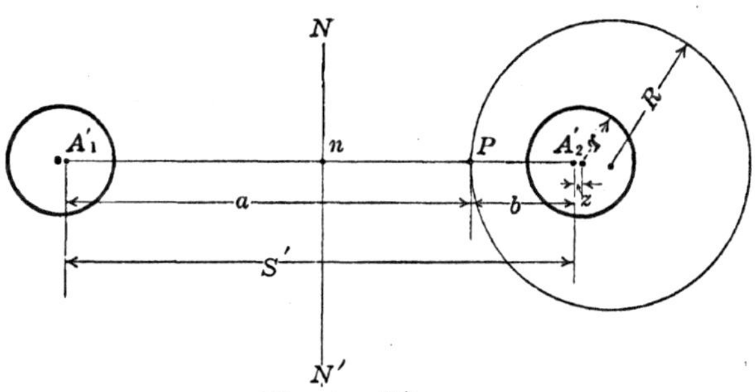

Fig. 186 (b)

Fig. 186.—Method of drawing lines of force and equipotential surfaces.

The equipotential surfaces are circles with centers on the line A'_1, A'_2. See equation (7), page 18. The distance of the center of the circles from A'_2 is $\frac{b^2}{a-b}$, and the radius is $\frac{ab}{a-b}$, where a and b must be so chosen that the permittances between circles are equal.

For any point p on the line $A'_1A'_2$ (Fig. 186*b*), the potential to the neutral plane due to A'_2 is

$$e_{2np} = \frac{-\psi}{2\pi Kk} \log_\epsilon \frac{2b}{S'}$$

Due to A'_1, it is

$$e_{1np} = \frac{\psi}{2\pi Kk} \log_\epsilon \frac{2a}{S'} = \frac{\psi}{2\pi Kk} \log_\epsilon \frac{2(S' - b)}{S'}$$

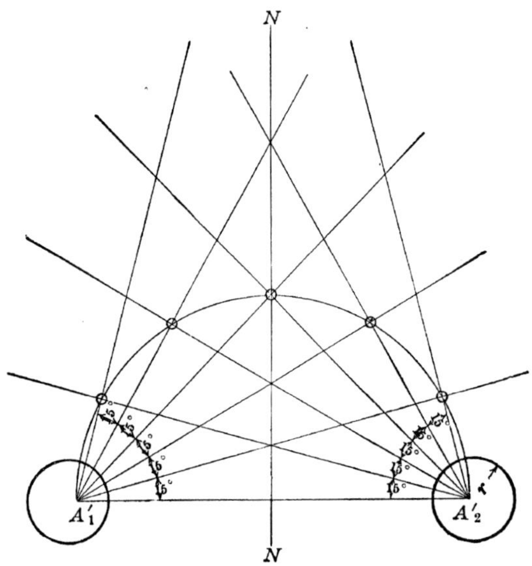

FIG. 187.—Graphical method of drawing lines of force between two cylinders.

The total voltage from any point p to the neutral is

$$e_{np} = e_{1np} + e_{2np} = \frac{\psi}{2\pi Kk}\left(\log_\epsilon \frac{2(S' - b)}{S'} - \log_\epsilon \frac{2b}{S'}\right)$$

$$e_{np} = \frac{\psi}{2\pi Kk} \log_\epsilon \frac{(S' - b)}{b}$$

It is desired to divide the field up into n equal voltages between the conductor surface and the neutral plane. The potential from the conductor to neutral is e_n and is known
Due to A'_2 :

$$e_{A2n} = \frac{-\psi}{2\pi Kk} \log_\epsilon \frac{2(r - z)}{S'}$$

Due to A'_1:

$$e_{A_1n} = \frac{\psi}{2\pi Kk} \log_\epsilon \frac{2(S' - (r - z))}{S'}$$

$$e_n = \frac{\psi}{2\pi Kk} \log \frac{S' - (r - z)}{r - z}$$

$$\therefore \psi = 2\pi Kk \frac{e_n}{\log_\epsilon \dfrac{S' - (r - z)}{r - z}}$$

Let β be the fraction of the voltage e_n it is desired to place between the neutral plane and any point p on the surface under consideration, then (Fig. 186*b*)

$$e_{np} = e_n\beta$$

$$\beta = \frac{1}{\log_\epsilon \dfrac{S' - (r - z)}{r - z}} \log_\epsilon \frac{(S' - b)}{b}$$

$$\log_\epsilon \frac{(S' - b)}{b} = \beta \log_\epsilon \frac{S' - (r - z)}{(r - z)} = \beta F$$

In this problem

$$\log_\epsilon \frac{S' - (r - z)}{r - z} = \text{constant} = F$$

$$F = \log_\epsilon \frac{9.8 - (1 - 0.1)}{(1 - 0.1)} = \log_\epsilon \frac{8.92}{0.92} = 2.290$$

$$\log_\epsilon \frac{S' - b}{b} = \beta F$$

$$\frac{S' - b}{b} = \text{anti-log } \beta F$$

$$b = \frac{S'}{(\text{anti-log } \beta F + 1)}$$

b is thus determined.

If it is desired to find b for the first circle from the conductor $\beta = 0.9$.

$$b_1 = \frac{9.84}{7.7 + 1} = \frac{9.84}{8.7} = 1.13$$

For the next circle

$$\beta = 0.8$$

$$b_2 = \frac{9.84}{6.23 + 1} = 1.36$$

15

Other values are found and tabulated in Table LXXXIII. a and R are thus found. The circles may now be drawn with radii R, centers on $A'_1A'_2$, and intersecting $A'_1A'_2$ b cm. from A'_2.

TABLE LXXXIII.—EQUIPOTENTIAL SURFACES

β	b		$a = S' - b$	$R = \frac{ab}{a - b}$	$a - b$	ab
1.0	b	0.900	8.90	1.00	8.00	8.00
0.9	b^1	1.110	8.69	1.27	7.58	9.65
0.8	b_2	1.353	8.45	1.62	7.09	11.49
0.7	b_3	1.648	8.15	2.07	6.50	13.45
0.6	b_4	1.98	7.82	2.65	5.84	15.46
0.5	b_5	2.38	7.42	3.50	5.04	17.65
0.4	b_6	2.81	7.09	4.65	4.28	19.90
0.3	b_7	3.29	6.51	6.65	3.22	21.40
0.2	b_8	3.80	6.00	10.35	2.20	22.80
0.1	b_9	4.34	5.46	21.15	1.12	23.67
0.0	b_{10}	4.90	4.90	∞	0.00	24.00

$F = 2.290 = \text{constant} \quad z = 0.1 \quad S' = S - 2.z = 9.8$

The gradient at any point, and therefore the equigradient surfaces may be found as follows:

The flux density at any point is

$$D = \frac{S'\psi}{2\pi x_1 x_2} \qquad \text{(Pages 21 and 23.)}$$

$$\psi = C_n e_n = \frac{2\pi k K e_n}{\log\left[\frac{S}{2r} + \sqrt{\left(\frac{S}{2r}\right)^2 - 1}\right]}$$

$$D = \frac{2S'\pi k K e_n}{2\pi x_1 x_2 \log_\epsilon\left[\frac{S}{2r} + \sqrt{\left(\frac{S}{2r}\right)^2 - 1}\right]}$$

$$g = \frac{D}{kK} = \frac{1}{x_1 x_2}\left[\frac{S'e_n}{\log_\epsilon\left[\frac{S}{2r} + \sqrt{\left(\frac{S}{2r}\right)^2 - 1}\right]}\right]$$

or for a given voltage, spacing and size of conductor, the term to the right is constant, and it follows:

$$g = \frac{1}{x_1 x_2} \text{ times a constant} = \frac{1}{x_1 x_2} M$$

is the gradient at any point x_1 cm. from A'_1 and x_2 cm. from A'_2. Putting x_1 in terms of x_2 and the angle α between x_2 and the line $A'_1A'_2$:

$$g = \frac{1}{x_2\sqrt{x_2^2 + S'^2 - 2S'x_2 \cos \alpha}} M$$

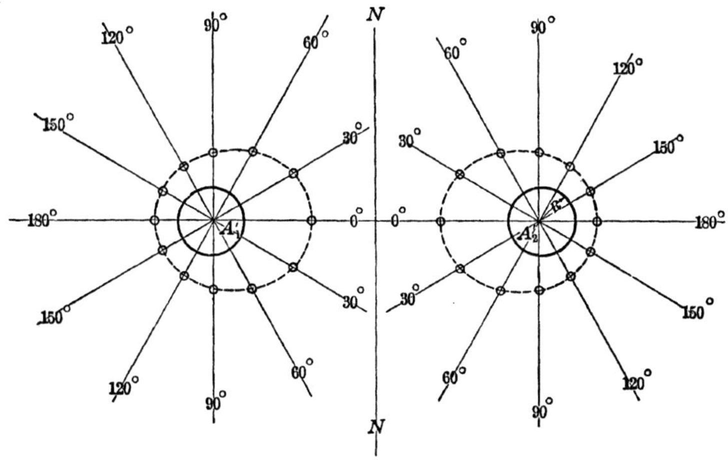

FIG. 188.—Method of plotting equigradient curves.

TABLE LXXXIV.—EQUIGRADIENT SURFACES

α		0°	30°	60°	90°	120°	150°	180°
$g = 21.4$	$x_2 =$	1.16	1.13	1.07	1.00	0.97	0.94	0.93
$g = 16.1$	$x_2 =$	1.64	1.60	1.44	1.32	1.27	1.23	1.20
$g = 10.7$	$x_2 =$	2.91	2.85	2.25	2.00	1.85	1.75	1.74
$g = 8.9$	$x_2 =$	4.90	3.40	2.71	2.37	2.19	2.06	2.02
$g = 8.0$	$x_2 =$	[1]	4.0	3.08	2.59	2.38	2.26	2.20
$g = 5.34$	$x_2 =$	$4.9 \pm 4\sqrt{-1}$[1]	8.20	4.70	3.81	3.27	3.17	3.10
$g = 2.67$	$x_2 =$	[1]	[1]	8.63	6.73	5.83	5.40	5.29

[1] Such a gradient does not exist on this line and hence x_2 is imaginary. (See plot Fig. 8, page 15.)

Gradient at Equidistant Points on the Conductor Surface

Angle bet. horizontal $A'_1A'_2$ and line through point and conductor center	0°	30°	60°	90°	120°	150°	180°
g	26.7	26.0	23.7	21.6	19.4	18.2	17.8

(1) Find the maximum gradient at the conductor surface for 100 kv. between conductors. This may be found directly from equation (12*a*), pages 24 and 29, and is 26.7 kv./cm.

(2) Find the gradient at six equidistant points on the conductor surface for 100 kv. between conductors. (See Table LXXXIV.)

(3) Calculate the equigradient curves for gradients of 21.4, 16.1, 10.7, 8.9, 8.0, 5.34, and 2.67 kv. per centimeter at 100 kv. between conductors. This may be done from the above equation by putting g equal to the required gradient, and finding x_2 for given values of α. The results are tabulated in Table LXXXIV, and method of plotting is shown in Fig. 188.

A complete plot of Case 9 is shown in Fig. 8, page 15. Note that in such a diagram, the permittance or elastance of each of the small cells bounded by sections of lines of force and equipotential surfaces is equal.

Case 10. Dielectric Fields in Three Dimensions.—The field of a conductor arrangement which must be considered in three dimensions, as a rod and torus, rod through a plane, etc., is generally represented on a plane figure in such a way that if the figure were revolved about its axis, the solid would be formed surrounded by its three-dimensional field. The small cells of the plane figure bounded by sections of lines of force and equipotential surfaces would form cells in the solid of equal permittance.

In the case of figures, as those for parallel wires, a wire in a cylinder, and parallel planes, it is possible to represent the field by considering only two dimensions. The height and thickness of the cells on the plane give constant permittance, or average $\frac{\text{height}}{\text{thickness}} = \text{constant}$. The third dimension is then the length of the wire and need not be considered in drawing these cells.

In the case of the three-dimensional field, the cells on the plane must be of such a height and thickness that the solid cells have constant permittance, or where the cell is small

$$\frac{hr}{t} = \text{constant}.$$

This can readily be seen from Fig. 189.

It is a general law that the cells must be so arranged that the stored energy may be a maximum or the permittance a maximum. In cases where the field need only be considered in two dimensions, it is thus possible, without great difficulty, to draw a field by a series of approximations in which the cells have a constant $\frac{h}{t}$ or add up to maximum permittance. This is also possible for a three-dimensional field, but extremely difficult because the cells must be drawn in such a way that the solid cells have constant permittance.

The best way in which to determine a field that cannot be readily calculated is experimentally. If the electrodes are immersed in an electrolyte in a large tank made of insulating material, and a small current passed between them, equipotential surfaces may be measured at equal voltage intervals on a plane through the axis of revolution by means of a galvanometer.[1] These surfaces correspond to the dielectric-equipotential surfaces. The lines of force may be drawn at right angles to these, so as to divide the field into solid cells of equal capacity (see above). It is difficult in practice to get results by this method when the problem includes several permittivities.

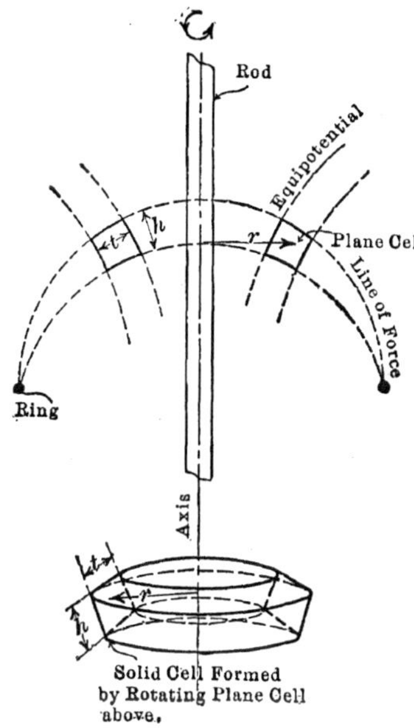

FIG. 189.—Dielectric field in three dimensions.

Theoretically it would be possible to use a solid material to represent, for instance, the porcelain shell of an insulator, and the electrolyte to represent the air. The resistivities of the two materials should then have the same ratio as the elastivities of porcelain and air. It is difficult to find a solid material with a resistivity in the order of that of an electrolyte.

As a less exact experimental method, the lines of force may be obtained by mica filings and the problem then solved by approximations. These methods may be developed into very useful ones for a study of flux control, etc.

Case 11. Effect of Ground on the Permittance and Gradient for Parallel Wires.—Such problems are solved by taking the "images" symmetrically below the ground.

[1] Fortesque has described this method. See A.I.E.E., March, 1913. Much more exact results may be obtained than those shown in this paper.

Voltages between AB due A, B, A_1, B_1 are

$$
\begin{aligned}
e_A &= +\left(\frac{\psi}{2\pi kK}\right)\log_e\frac{S}{r}\\
e_B &= -\left(\frac{\psi}{2\pi kK}\right)\log_e\frac{r}{S}\\
e_{A_1} &= -\left(\frac{\psi}{2\pi kK}\right)\log_e\frac{a}{2h}\\
e_{B_1} &= +\left(\frac{\psi}{2\pi kK}\right)\log_e\frac{2h}{a}
\end{aligned}
\quad \left(\begin{matrix}S \cong S' \text{ where the wires}\\ \text{are far apart.}\end{matrix}\right)
$$

The total voltage is

$$e = e_A + B_B + e_{A_1} + e_{B_1} = \frac{\psi}{\pi kK}\left(\log\frac{S}{r} + \log_e\frac{2h}{a}\right) = \frac{\psi}{\pi kK}\log\frac{S}{r}\frac{2h}{a}$$

$$C = \frac{\psi}{e} = \frac{\pi kK}{\log\frac{S}{r}\frac{2h}{a}} \text{ (Bet. lines)}$$

Where the wires are far apart the gradient is

$$g = \frac{e}{2\,r\log_e\frac{S}{r}\frac{2h}{a}}$$

The problem for a number of wires may be solved in the same way. The fluxes from the different wires may then not be the same and the solution is more difficult as a number of simultaneous equations must be written and solved.

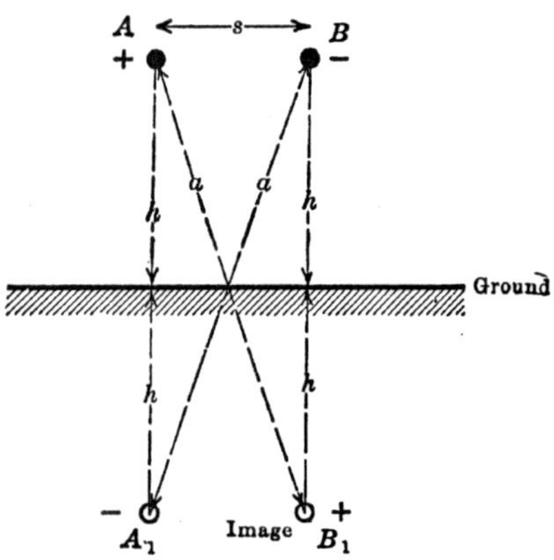

FIG. 190.—Effect of ground on capacity between parallel wires.

Case 12. Three-phase Dielectric Field with Symmetrical and Unsymmetrical Spacings.—The fluxes between conductors on a three-phase line vary sinusoidally with the voltages. The instantaneous values of voltages may be added algebraically as above. The effective or maximum values are found by geometrical addition.

To illustrate: find the fluxes for three three-phase conductors in an equilateral triangle, and also for flat spacing. In order to greatly simplify the problem, the effects of ground or "images" will be neglected, and the conductors considered far apart.

Then due to fluxes from

$$\qquad\qquad A \qquad\quad | \qquad\quad B \qquad\quad | \qquad\quad C$$

$$(a)\ E_{AB} = \frac{1}{2\pi kK}\left(\psi_A \log_e \frac{AB}{r} + \psi_B \log_e \frac{r}{BA} + \psi_C \log_e \frac{CB}{CA}\right) = e \sin \theta$$

$$(b)\ E_{BC} = \frac{1}{2\pi kK}\left(\psi_A \log_e \frac{AC}{AB} + \psi_B \log_e \frac{BC}{r} + \psi_C \log_e \frac{r}{CB}\right) = e \sin (\theta - 120)$$

Only two of the three equations which may be written as above are independent, since the sum of the voltages must be zero. The other independent equation is

$$(c) \qquad \psi_A + \psi_B + \psi_C = 0$$

Conductors Spaced in a Triangle (A at the apex, B and C at the base) (equilateral triangle).

Substituting spacing S in (a) and (b) and solving for ψ_A, ψ_B, and ψ_C

$$\psi_B = -\frac{2\pi kKe}{3 \log_e \frac{S}{r}} (1.5 \sin \theta + 0.866 \cos \theta)$$

Let $\quad e(1.5 \sin \theta + 0.866 \cos \theta) = e_a (\sin \theta - \alpha)$

put $\quad \theta = 90$ and $\theta = 0$ and solve for e_a and α

$$\psi_B = \frac{2\pi kKe}{\sqrt{3} \log_e \frac{S}{r}} \sin (\theta - 150) = \frac{1.16(\pi kK)e}{\log_e \frac{S}{r}} \sin (\theta - 150)$$

$$\psi_A = \frac{2\pi kKe}{\sqrt{3} \log_e \frac{S}{r}} \sin (\theta - 30) = \frac{1.16(\pi kK)e}{\log_e \frac{S}{r}} \sin (\theta - 30)$$

ψ_C may be found in the same way, or for this particular case, ψ_A, ψ_B, and ψ_C are equal by symmetry.

For single phase:

$$\psi = \frac{2\pi kKe}{2 \log_e \frac{S}{r}}$$

Therefore, when the wires are far apart, and with the same voltage between lines, the three-phase stress is $\frac{2}{\sqrt{3}}$ times the single-phase stress.

Flat Spacing $\underset{\circ}{A}\ \underset{\circ}{B}\ \underset{\circ}{C}$.—Putting S, S, and $2S$ in a and b and solving as before,

$$\psi_B = \frac{2\pi k K e}{\sqrt{3}\ \log_e \dfrac{S}{r} - 0.58 \log 2} \sin(\theta - 150)$$

$$= \frac{2\pi k K\, e}{\sqrt{3}\ \log_e \dfrac{S}{r} - 0.4} \sin(\theta - 150)$$

The flux, and therefore the stress, when the wires are far apart, is greatest on the middle wire. For $\frac{S}{r} = 500$ it is 4 per cent. greater than on the wires with the same S and triangular spacing as above. ψ_A, and ψ_C are 6 per cent. lower than for the triangular spacing. The gradients vary in the same way. Corona, therefore, starts on the center wire at a 4 per cent. lower voltage, and on the outer wires at a 6 per cent. higher voltage than for triangular spacings.

Case 13. Occluded Air in Insulation.—It is interesting to estimate the effect of occluded air in solid insulation. Assume that in the process of manufacture air bubbles have formed in a sheet of rubber insulation. The sheet is 1 cm. thick. The "bubbles" are thin compared to the rubber, and long in the direction of the length of the sheet. It is estimated that the largest ones are 0.01 cm. thick, and 0.1 cm. long and wide. The electrodes which the rubber insulates may be assumed as being practically parallel planes. The working voltage is 40 kv., or the stress is 40 kv./cm. effective in the rubber. As the air bubbles are not thick enough to greatly disturb the field, the same flux passes through the air as through the rubber. The permittivity of the rubber is 3. The stress on the air is, therefore, $3 \times 40 = 120$ kv./cm. effective. Air breaks down at 21.2 kv./cm. effective at atmospheric pressure. It seems probable that these bubbles will break down, even after allowance is made for the extra strength of thin films, and a possible pressure higher than atmospheric. It is probable that the solid insulation would soon break down on account of heat and chemical action.

Case 14. General.—(*a*) Estimate the visual corona voltage when wires are wet. Compare with the visual corona voltage when wires are dry.

Calculate g_v from the formula on page 67, Chapter III. Insert the value in formula (20). Maximum e_v to neutral is thus found. If the voltage used is a sine wave, reduce to effective kv. by dividing by $\sqrt{2}$. For a three-phase line the voltage between wires may be found by multiplying by $\sqrt{3}$; for a single-phase line, by multiplying by 2. Compare with dry visual critical voltage calculated from equation (20); page 43.

(*b*) At what voltage will the above wires spark over wet and dry single phase; three phase?

Estimate dry spark-over voltage from equation given on page 83, Chapter IV. Estimate wet arc-over voltage by assuming needle gap spark-over.

(*c*) Calculate the dry arc-over curve for a 10-cm. sphere (grounded) at $\delta = 0.90$, and spacings from 1.5 to 10 cm.

Use equation (13*b*), Chapter IV. Estimate a wet spark-over curve as outlined for spheres on page 105, Chapter IV.

(*d*) What is the voltage required to puncture 0.5 cm. of paper insulation when the time of application is limited to 1/120000 second? In 100 seconds?

Use equation on page 179, Chapter VII, of the form

$$g_s = g\left(1 + \frac{a}{\sqrt[4]{T}}\right)$$

$$e_s = g_s \times \text{thickness.}$$

(*e*) Estimate the loss per cubic centimeter at 1000 cycles in a piece of varnished cambric, at 5.0 kv./mm., 25 deg. C. Use equation page 185, Chapter VII.

(*f*) What is the breakdown gradient of a piece of porcelain 2 cm. thick?

$$g = 7.5\left(1 + \frac{0.94}{\sqrt{t}}\right)$$

where t = thickness in mm.
g = gradient in kv./mm. (eff.)

(See Chapter VII, page 174.)

DATA APPENDIX

MEASURED CORONA LOSS

Indoor Line—60-cycle

The current and watts given are measured values due to corona, divided by the total conductor length in kilometers. Corrections have been made for transformer and leads. The voltage is given to neutral. As these measurements were made on a single-phase line, the voltages between wires were twice the value given.

CORONA LOSS—INDOOR LINE—60-CYCLE

Test 10B			Test 11B		
Eff. kv. to neutral, e_n	Amp. per km.	Loss kw./km., p	Eff. kv. to neutral, e_n	Amp. per km.	Loss kw./km., p
10.52	....	0.07	13.7	....	0.33
12.52	....	0.51	16.7	....	0.92
14.80	....	1.40	18.1	....	1.11
17.10	....	3.20	19.9	0.070	1.64
18.20	....	4.02	24.7	0.100	3.95
19.60	....	6.83	29.3	0.150	6.90
22.30	0.225	9.44	33.2	0.189	10.40
24.90	0.310	15.03	36.3	0.223	14.20
27.20	0.395	20.63	39.7	0.268	18.75
29.70	....	....	44.0	0.325	26.73
28.00	0.404	22.40	47.3	0.380	34.20
25.90	0.342	16.80	50.2	....	....
23.60	0.263	11.80	45.2	0.350	29.20
20.80	....	7.16	41.5	0.293	22.20
18.40	....	4.14	37.4	0.238	15.69
16.10	....	1.98	31.0	0.163	8.08
			27.0	0.128	5.29
			22.3	0.082	2.53

Spacing, 15.25 cm.
Radius, 0.032 cm.
Total cond. length, 0.0838 km.
δ = 1.02.

Spacing, 30.5 cm.
Radius, 0.032 cm.
Total cond. length, 0.0838 km.
δ = 1.02.

Test 13*B*			Test 15*B*		
Eff. kv. to neutral, e_n	Amp. per km.	Loss kw./km., p	Eff. kv. to neutral, e_n	Amp. per km.	Loss kw./km., p
19.5	0.051	0.96	17.3	....	0.35
24.3	0.071	1.78	23.7	0.059	1.17
29.6	0.099	3.23	30.2	0.078	2.34
34.0	0.133	4.95	36.1	0.109	3.92
38.0	0.146	6.49	42.2	....	5.76
43.0	....	9.13	49.8	0.166	9.26
46.7	0.196	12.51	56.7	0.200	13.21
52.4	0.224	16.35	62.5	0.223	16.70
57.5	....	21.00	66.5	0.246	20.00
62.5	0.285	26.25	71.4	....	24.35
68.3	0.327	34.40	78.6	0.310	30.95
73.1	0.359	40.60	84.4	0.322	37.20
79.1	0.385	50.50	89.4	0.345	41.90
84.0	....	59.50	95.4	....	50.50
90.0	0.430	72.00	99.3	0.395	58.60
93.6	0.460	83.10	91.0	0.369	45.60
102.0	0.510	93.40	81.6	0.310	34.10
			67.9	....	21.55
			52.1	0.177	10.46
			39.1	0.126	4.44
			27.0	0.068	1.74

Spacing, 61 cm.
Radius, 0.032 cm.
Total cond. length, 0.0838 km.
$\delta = 1.03$.

Spacing, 91.5 cm.
Radius, 0.032 cm.
Total cond. length, 0.0838 km.
$\delta = 1.012$.

Test 20B			Test 21B		
Eff. kv. to neutral, e_n	Amp. per km.	Loss kw./km., p	Eff. kv. to neutral, e_n	Amp. per km.	Loss kw./km., p
17.8	....	0.21	19.8	....	0.81
24.6	....	1.29	24.5	....	1.15
30.1	....	2.17	27.2	....	2.02
36.2	0.110	3.71	30.7	....	2.19
42.0	0.128	5.16	34.8	....	3.42
47.0	0.159	6.94	39.8	0.129	3.86
53.2	0.190	9.80	44.8	0.156	5.04
62.0	0.233	14.60	50.4	0.177	6.47
67.6	0.258	19.23	55.1	0.212	8.45
73.4	0.273	21.12	60.6	0.234	10.80
78.1	....	26.70	64.6	0.246	12.40
85.2	0.346	33.10	70.0	0.266	15.10
93.2	0.385	42.70	77.9	0.310	19.30
100.2	....	51.70	87.6	....	26.60
87.0	0.350	35.60	94.3	....	32.60
76.1	0.298	25.00	101.5	....	41.00
65.2	0.250	15.95			
49.4	0.178	7.44			
40.1	0.114	4.55			

Spacing, 122 cm.
Radius, 0.032 cm.
Total cond. length, 0.0421 km.
δ = 1.018.

Spacing, 183 cm.
Radius, 0.032 cm.
Total cond. length, 0.0342 km.
δ = 1.012.

Test 22*B*			Test 24*B*		
Eff. kv. to neutral, e_n	Amp. per km.	Loss kw./km., p	Eff. kv. to neutral, e_n	Amp. per km.	Loss kw./km., p
20.7	....	0.72	20.1	....	0.24
24.5	....	1.01	25.3	....	0.98
27.2	....	1.21	32.0	....	1.84
30.9	....	1.98	36.0	0.103	3.18
34.6	0.100	2.34	41.0	0.120	4.52
40.1	0.121	3.57	45.7	0.138	6.11
45.2	0.149	4.53	50.3	....	8.30
50.3	0.171	5.71	57.0	0.185	11.57
55.1	....	7.43	62.2	0.210	15.10
60.2	0.215	9.13	67.3	0.240	19.15
64.6	0.230	11.05	72.4	0.266	23.20
69.5	....	11.99	77.5	0.290	27.50
78.5	....	16.37	82.3	....	32.10
87.5	0.345	23.00	87.8	....	37.90
94.2	0.380	26.00	92.5	....	43.80
103.2	0.417	34.50	97.3	....	50.80
			100.5	....	55.60
			79.0	....	28.70
			65.0	....	17.10
			52.7	....	9.90

Test 22*B*:
Spacing, 274.5 cm.
Radius, 0.032 cm.
Total cond. length, 0.0342 km.
$\delta = 1.012$.

Test 24*B*:
Spacing, 91.5 cm.
Radius, 0.057 cm.
Total cond. length, 0.0818 km.
$\delta = 1.0009$.

Test 26*B*			Test 28*B*		
Eff. kv. to neutral, e_n	Amp. per km.	Loss kw./km., p	Eff. kv. to neutral, e_n	Amp. per km.	Loss kw./km., p
20.0	....	0.37	10.6	....	....
23.2	0.053	0.98	14.2	....	....
29.0	....	2.24	16.1	....	0.18
32.8	....	3.36	18.4	....	0.43
37.1	0.124	5.14	20.2	....	0.98
41.1	0.141	6.60	22.2	0.061	1.34
46.7	0.173	9.90	24.0	0.069	1.83
51.3	0.202	12.32	25.0	0.072	2.20
57.1	0.232	18.30	27.0	0.087	2.93
61.1	0.262	22.30	30.2	0.120	5.00
66.2	0.290	28.40	34.0	0.157	7.88
72.0	0.332	36.60	36.7	0.178	10.30
75.7	0.334	40.80	36.7	0.181	10.10
80.7	....	50.50	40.7	0.212	14.80
85.0	0.413	56.50	44.5	0.265	20.30
88.7	0.431	64.00	47.1	0.298	24.80
91.0	0.464	71.40	51.7	0.369	35.20
96.5	0.515	83.40	49.6	0.342	30.80
101.0	0.562	104.00	43.0	0.251	18.20
82.8	0.396	54.50			
74.7	0.333	39.00			
69.7	....	33.00			
60.0	0.255	22.70			

Spacing, 0.61 cm.
Radius, 0.057 cm.
Total cond. length, 0.08186 km.
$\delta = 1.002$.

Spacing, 30.5 cm.
Radius, 0.057 cm.
Total cond. length, 0.0818 km.
$\delta = 0.993$.

Test 30*B*			Test 32*B*		
Eff. kv. to neutral, e_n	Amp. per km.	Loss kw./km., p	Eff. kv. to neutral, e_n	Amp. per km.	Loss kw./km., p
21.7	....	0.18	22.5	....	0.21
25.6	....	0.92	27.2	0.049	0.31
31.5	....	2.45	32.5	0.062	1.41
36.2	....	4.30	37.2	0.086	2.70
40.5	0.138	5.80	41.1	0.100	3.68
45.7	0.156	8.60	45.6	0.123	5.03
49.7	0.180	11.70	50.3	0.146	7.35
54.2	0.204	15.30	55.0	0.178	10.18
59.5	0.238	19.50	60.0	0.194	13.50
64.0	0.269	24.60	65.7	0.222	17.40
68.2	0.290	29.70	72.0	0.257	23.20
73.0	0.322	36.30	77.2	0.278	27.60
78.5	0.352	45.20	82.5	0.298	31.40
83.5	0.384	54.00	89.2	....	40.00
88.2	0.441	63.00	92.2	....	44.00
95.0	0.486	76.00	85.2	....	35.80
96.5	0.500	81.20	70.0	....	20.80
100.0	0.530	92.00	61.7	....	14.32
103.0	0.600	114.30			
90.2	0.459	60.20			
79.5	0.356	45.30			
75.2	0.319	38.60			
65.5	....	26.20			
51.7	....	13.40			

Spacing, 61 cm.
Radius, 0.071 cm.
Total cond. length, 0.0815 km.
δ = 0.98.

Spacing, 91.5 cm.
Radius, 0.914 cm.
Total cond. length, 0.0815 km.
δ = 1.002.

Test 33*B*			Test 41*B*		
Eff. kv. to neutral, e_n	Amp. per km.	Loss kw./km., p	Eff. kv. to neutral, e_n	Amp. per km.	Loss kw./km., p
24.7	....	0.31	28.3	....	....
29.8	....	1.53	35.5	0.071	2.72
32.6	....	2.45	41.0	0.108	4.90
35.6	....	3.50	48.6	0.184	10.40
39.3	....	4.96	53.7	0.208	13.25
43.7	....	6.37	59.0	0.234	17.50
47.5	....	9.57	63.8	0.265	23.40
51.5	....	12.9	69.4	0.306	30.02
56.3	....	15.5	73.6	0.353	35.70
60.2	....	20.8	71.9	0.325	34.10
65.5	....	34.8	60.1	0.244	19.85
67.0	....	29.6	54.2	0.210	13.65
54.4	....	15.3	59.9	0.250	19.35
			57.0	0.224	16.30
			47.0	0.161	8.87

Spacing, 61 cm.
Radius, 0.0914 cm.
Total cond. length, 0.0815 km.
δ = 1.006.

Spacing, 61 cm.
Radius, 0.105 cm.
Total cond. length, 0.0423 km.
δ = 1.001.

Test 45*B*			Test 47*B*		
Eff. kv. to neutral, e_n	Amp. per km.	Loss kw./km., p	Eff. kv. to neutral, e_n	Amp. per km.	Loss kw./km., p
22.3	....	0.06	45.0	0.080	0.37
25.8	....	0.12	50.0	0.100	0.74
29.7	....	0.34	54.5	0.104	1.90
35.7	....	1.41	60.5	....	5.76
39.5	....	2.70	65.7	0.152	9.13
44.2	....	4.78	70.5	0.174	12.60
50.2	....	8.10	75.5	0.205	16.30
54.0	....	10.70	79.0	0.230	18.75
59.5	....	14.70	87.0	....	26.80
64.5	....	19.50	91.7	....	31.40
69.0	....	24.80	96.5	....	38.20
74.0	....	30.10	93.5	0.321	34.10
79.7	....	37.20	90.0	....	29.80
82.7	....	43.40	82.0	....	21.70
88.7	....	56.20	75.8	....	16.20
98.0	....	59.00	58.0	....	4.22
99.7	....	80.50	42.6	....	0.49
103.0	....	84.50	40.0	....	0.24
86.7	....	52.20			
75.5	....	32.70			

Spacing, 61 cm.
Radius, 0.164 cm.
Total cond. length, 0.0185 km.
$\delta = 0.996$.

Spacing, 91.5 cm.
Radius, 0.256 cm.
Total cond. length, 0.0815 km.
$\delta = 0.996$.

Test 48*B*			Test 49*B*		
Eff. kv. to neutral, e_n	Amp. per km.	Loss kw./km., *p*	Eff. kv. to neutral, e_n	Amp. per km.	Loss kw./km., *p*
44.1	....	0.61	40.5	....	1.16
47.2	....	1.11	44.5	....	3.13
51.7	....	3.87	43.0	....	2.27
54.5	....	5.90	46.8	....	5.70
60.8	....	10.60	50.6	....	12.30
67.2	....	17.15	55.6	....	19.25
71.5	....	20.70	58.8	....	....
75.5	....	26.70	53.8	....	16.05
81.7	....	33.30	49.6	....	9.70
76.6	....	27.10	45.7	....	5.27
78.5	....	29.40	43.2	....	2.82
74.0	....	23.90	39.3	....	1.23
69.5	....	18.30			
68.0	....	16.55			
58.2	....	8.10			
52.7	....	4.10			

Spacing, 61 cm.
Radius, 0.256 cm.
Total cond. length, 0.0815 km.
$\delta = 1.00$.

Spacing, 30.5 cm.
Radius, 0.256 cm.
Total cond. length, 0.0815 km.
$\delta = 0.996$.

Test 51B			Test 54B		
Eff. kv. to neutral, e_n	Amp. per km.	Loss kw./km., p	Eff. kv. to neutral, e_n	Amp. per km.	Loss kw./km., p
62.5	0.159	9.15	25.5	0.051	....
68.0	0.190	14.55	31.8	....	....
72.0	0.220	20.10	37.7	....	0.06
78.0	0.261	28.45	42.7	....	0.30
87.3	0.318	40.20	47.1	....	0.30
91.5	....	49.70	57.0	....	0.49
82.0	....	34.60	63.2	....	1.03
72.8	....	22.10	69.7	....	1.81
56.0	....	5.65	76.5	0.152	5.08
61.5	....	10.20	78.7	0.164	6.48
67.2	....	15.75	84.5	0.184	12.22
70.2	....	19.30	89.3	0.210	17.20
			95.0	0.238	23.90
			99.1	0.263	28.40
			102.2	....	31.90
			97.2	....	26.00
			92.0	....	20.90
			87.0	....	16.50
			76.7	....	5.82
			82.0	....	9.10
			72.1	....	2.67

Spacing, 61 cm.
Radius, 0.333 cm.
Total cond. length, 0.0817 km.
$\delta = 0.999$.

Spacing, 91.5 cm.
Radius, 0.464 cm.
Total cond. length, 0.0825 km.
$\delta = 0.982$.

All of the above tests were taken at a temperature of about 25 deg. C.

MEASURED CORONA LOSS

Outdoor Line—60 cycle

Columns 1, 2, and 3 are actual measured values and include transformer and lead losses. Column 4, the actual corona loss, for the length of line used in the test is obtained from Column 3 by subtracting transformer and lead losses.

These tests were made on comparatively long single-phase lines out of doors, and the conductor surfaces, etc., were not in as good condition as in the case of the indoor line. Transformer losses for several temperatures are given. The voltage values are effective between lines.

Test No. 146, Line *A*[1]			
Kv. bet. lines	Amp.	Kw.	Kw. line loss, *p*
63.5	0.056	0.07	0.01
80.5	0.077	0.12	0.02
90.1	0.092	0.15	0.02
107.5	0.113	0.30	0.12
115.2	0.121	0.35	0.14
126.2	0.135	0.63	0.37
134.2	0.146	0.85	0.55
142.5	0.154	1.29	0.95
150.0	0.164	1.95	1.45
158.0	0.173	2.69	2.25
166.1	0.185	4.00	3.48
165.0	0.183	3.51	3.02
173.7	0.196	5.00	4.45
163.4	0.184	2.70	2.23
170.4	0.193	4.20	3.67
181.0	0.198	6.06	5.42
203.0	0.251	12.84	12.04
199.2	0.243	11.50	10.73
193.4	0.227	9.10	8.49
176.4	0.197	4.74	4.17
165.6	0.184	2.70	2.21
162.8	0.180	2.38	1.81
154.4	0.169	1.28	0.85
166.0	0.176	2.94	2.45
184.4	0.198	6.49	5.87
172.0	0.193	3.85	3.31
160.0	0.177	2.03	1.57
146.2	0.162	0.80	0.42
138.0	0.150	0.52	0.19
127.6	0.137	0.40	0.12
122.5	0.129	0.33	0.08
111.2	0.117	0.25	0.04
101.0	0.105	0.20	0.03

Test No. 18, Line *A*			
Kv. bet. lines	Amp.	Kw.	Kw. line loss, *p*
80.0	0.040	0.12	0.01
90.0	0.100	0.16	0.02
101.1	0.107	0.20	0.04
112.0	0.113	0.25	0.05
121.6	0.123	0.30	0.06
129.5	0.131	0.35	0.07
140.0	0.146	0.49	0.16
150.0	0.160	0.76	0.38
160.0	0.172	1.60	1.17
152.0	0.162	0.90	0.51
164.2	0.174	2.00	1.55
172.0	0.187	3.40	2.90
183.2	0.205	5.60	5.02
188.2	0.210	6.92	6.30
196.4	0.223	9.02	8.42
202.2	0.237	11.06	10.36
206.0	0.242	12.90	12.09
187.2	0.211	6.95	6.34
196.4	0.225	9.60	8.90

Total conductor length, 109,500 cm.
Spacing, 310 cm.
No. 3/0 7-strand hard-drawn copper-weathered cable, diam. 1.18 cm.
Temperature, wet, 16 deg. C.
dry, 18.5.
Barometer, 75.5 cm.
Bright sun, wind.

Test No. 146, Line *A*

Total conductor length, 109,500 cm.
Spacing, 310 cm.
No. 3/0 7-strand cable, dia. 1.18 cm.
Temperature, wet, 24 deg. C.
dry, 30 deg. C.
Barometer, 75.7 cm.
Hazy.

[1] This curve was taken after the line had been standing idle over a month in the summer. The "going up" points show an excess loss due to dust and dirt on the conductor. This disappears at high voltage and does not show in the "coming down" readings.

Test No. 102			
Kv. bet. lines	Amp.	Kw.	Kw. line loss, p
201.0	0.232	6.05	6.65
211.0	0.277	9.10	8.63
189.0	0.210	3.54	3.19
181.8	0.200	2.36	2.04
170.8	0.189	1.10	0.80
160.0	0.176	0.60	0.36
149.0	0.162	0.36	0.16
201.0	0.231	6.15	5.75
149.0	0.162	0.39	0.19
140.5	0.150	0.29	0.12
135.5	0.145	0.25	0.10
124.5	0.131	0.20	0.08
113.5	0.118	0.16	0.08
102.3	0.103	0.13	0.07

Total conductor length, 109,500 cm.
Spacing, 310 cm.
No. 3/0 7-strand H. D. copper-weathered cable, diam. 1.18 cm.
Temperature, wet, 1
dry, 1
Barometer, 7.47 cm.
Cloudy.

Test No. 84, Line *A*			
Kv. bet. lines	Amp.	Kw.	Kw. line loss, *p*
120.0	0.138	0.24	0.15
129.0	0.150	0.30	0.19
160.0	0.175	0.78	0.61
181.0	0.202	3.65	3.40
189.0	0.212	4.65	4.36
203.0	0.237	7.84	7.48
213.0	0.252	11.20	10.78
205.0	0.239	8.70	8.18

Total conductor length, 108,500 cm.
Spacing, 310 cm.
No. 3/0 7-strand cable (H. D. copper-weathered), 1.18 cm.
Temperature, wet, 1 deg. C.
dry, 3 deg. C.
Barometer, 75.2 cm.
Cloudy.

Test No. 105, Line *A*			
Kv. bet. lines	Amp.	Kw.	Kw. line loss, *p*
79.8	0.080	0.03	0.01
90.7	0.093	0.04	0.01
101.5	0.106	0.06	0.02
109.5	0.114	0.08	0.03
120.5	0.127	0.10	0.04
130.0	0.139	0.14	0.06
141.5	0.154	0.19	0.09
147.0	0.165	0.21	0.08
153.6	0.168	0.25	0.12
159.0	0.178	0.30	0.16
169.8	0.199	0.51	0.35
174.0	0.190	0.70	0.53
181.0	0.198	1.20	1.02
186.2	0.204	1.74	1.55
192.6	0.212	2.70	2.49
200.6	0.221	4.00	3.77
208.6	0.237	5.60	5.34
216.0	0.247	7.40	7.13
221.0	0.259	9.00	8.70
227.0	0.271	11.00	10.66
234.0	0.288	13.60	13.25
189.0	0.210	2.30	2.10
195.0	0.217	3.10	2.88
203.8	0.229	4.96	4.72
212.0	0.242	6.70	6.44
219.0	0.257	8.60	8.31

Total conductor length, 109,500 cm.
Spacing, 310 cm.
No. 3/0 7-strand H. D. copper-weathered cable, diam. 1.18 cm.
Temperature, wet, 13 deg. C.
dry, 13 deg. C.
Barometer, 76.2 cm.
Bright sun, no wind, snow on ground.

Test No. 100, Line *B*			
Kv. bet. lines	Amp.	Kw.	Kw. line loss, *p*
67.0	....	0.02	0.02
77.0	0.025	0.03	0.02
88.0	0.028	0.05	0.02
98.9	0.035	0.07	0.03
109.5	0.040	0.12	0.07
119.5	0.043	0.22	0.14
128.0	0.050	0.42	0.32
137.0	0.054	0.90	0.78
144.0	0.060	1.94	1.80
161.2	0.078	4.50	4.31
153.0	0.070	3.04	2.88
173.8	0.090	6.60	6.47
185.0	0.103	8.72	8.36
200.0	0.106	11.90	11.59
185.0	0.103	8.76	8.50
159.0	0.078	4.10	3.92
139.0	0.058	1.22	1.10
161.2	0.080	4.70	4.51
211.8	0.135	14.80	14.46

Total conductor length, 29,050 cm.
Spacing, 91.4 cm.
0.375-in. galv. steel cable, diam. 0.953 cm.
Temperature wet, 1 deg. C.
dry, 1 deg. C.
Barometer, 74.7 cm.
Cloudy.

Test No. 73, Line *B*			
Kv. bet. lines	Amp.	Kw.	Kw. line loss, *p*
43.0	0.016	....	....
60.0	0.022	....	....
69.7	0.026	0.08	0.06
80.6	0.030	0.10	0.07
90.5	0.034	0.15	0.11
101.5	0.038	0.30	0.26
91.0	0.034	0.09	0.05
90.5	0.034	0.10	0.06
70.3	0.026	0.06	0.04
101.6	0.038	0.17	0.12
101.6	0.038	0.17	0.12
109.5	0.041	0.40	0.36
105.5	0.040	0.14	0.09
115.0	0.040	0.14	0.09
115.0	0.0425	0.88	0.82
121.5	0.048	0.16	0.09
126.5	0.053	2.00	1.93
130.5	0.055	2.48	2.40
140.5	0.064	3.70	3.61
144.5	0.073	4.26	4.16
70.5	....	0.03	0.00
91.5	0.030	0.06	0.02
106.0	0.038	0.18	0.13
150.0	0.078	4.80	4.69
156.4	0.083	5.80	5.68
161.0	0.089	6.72	6.67
166.0	0.093	7.50	7.36

Total conductor length, 29,050 cm.
Spacing, 91.4 cm.
0.23-in. galv. steel cable, diam. 0.585 cm.
Temperature wet, 1 deg. C.
dry, 3 deg. C.
Barometer, 75.2 cm.
Cloudy.

Test No. 79, Line *B*				Test No. 80, Line *B*			
Kv. bet. lines	Amp.	Kw.	Kw. line loss, *p*	Kv. bet. lines	Amp.	Kw.	Kw. line loss, *p*
213.0	0.105	8.64	8.38	81.0	0.029	0.07	0.04
205.0	0.010	7.68	7.40	91.0	0.032	0.09	0.05
202.0	0.094	7.40	7.13	100.5	0.035	0.12	0.07
186.0	0.088	6.00	5.80	110.5	0.038	0.16	0.11
181.0	0.081	5.00	4.80	120.5	0.041	0.40	0.34
168.4	0.072	3.96	3.81	130.5	0.048	1.30	1.22
159.6	0.063	3.00	2.88	139.5	0.055	2.25	2.17
150.0	0.058	2.24	2.13	153.0	0.067	3.20	3.09
138.0	0.048	1.14	1.06	160.0	0.075	4.40	4.28
120.0	0.043	0.20	0.14	172.0	0.084	5.70	5.54
120.0	....	0.26	0.20	181.0	0.094	7.00	6.82
110.0	0.071	0.19	0.14	192.0	0.103	8.50	8.28
99.0	0.068	0.13	0.09	199.0	0.109	9.40	9.15
				213.0	0.124	11.70	11.38

Test No. 79: Total conductor length, 29,050 cm.
Spacing, 244 cm.
0.23-in. galv. steel cable, diam. 0.585 cm.
Temperature, wet, 1 deg. C.
dry, 3 deg. C.
Barometer, 72.5 cm.
Cloudy.

Test No. 80: Total conductor length, 29,050 cm.
Spacing, 152 cm.
0.23-in. galv. steel cable, diam. 0.585 cm.
Temperature, wet, 1 deg. C.
dry, 3 deg. C.
Barometer, 72.5 cm.
Cloudy.

CORONA LOSS—OUTDOOR LINE—60-CYCLE

Test No. 125, Line *B*				Test No. 126, Line *B*			
Kv. bet. lines	Amp.	Kw.	Kw. line loss, *p*	Kv. bet. lines	Amp.	Kw.	Kw. line loss, *p*
80.0	0.025	0.06	0.05	100.0	0.031	0.12	0.09
88.0	0.031	0.13	0.11	110.0	0.037	0.22	0.17
101.0	0.037	0.32	0.29	119.0	0.041	0.44	0.36
110.0	0.041	0.74	0.68	131.0	0.050	1.36	1.22
120.0	0.050	1.67	1.59	142.0	0.056	2.38	2.16
128.0	0.056	2.56	2.44	151.0	0.065	3.23	2.92
140.0	0.067	4.00	3.80	160.0	0.074	4.20	3.78
150.0	0.08	5.42	5.12	171.0	0.082	5.45	4.93
159.6	0.09	6.86	6.46	181.0	0.09	6.56	5.89
168.4	0.101	8.30	7.80	194.0	0.102	8.20	7.34
181.0	0.112	10.36	9.68	202.0	0.111	9.26	8.30
190.0	0.122	12.24	11.44	212.0	0.117	10.84	9.74
201.0	0.134	14.68	13.61	222.0	0.128	12.44	11.18
213.0	0.148	17.28	16.14	231.0	0.135	13.80	12.38
206.0	0.144	15.76	14.72	225.0	0.129	12.88	11.58
196.6	0.128	13.60	12.70	217.0	0.124	11.64	10.45
186.2	0.117	11.44	10.69	205.0	0.112	9.70	8.70
175.0	0.103	9.20	8.59	196.6	0.104	8.56	7.68
165.6	0.096	7.76	7.27	186.6	0.096	7.32	6.56
153.4	0.083	5.92	5.57	176.0	0.086	6.12	5.51
143.0	0.074	4.56	4.33	165.0	0.078	4.96	4.49
134.0	0.064	3.34	3.18	156.4	0.069	3.96	3.59
123.0	0.053	2.00	1.90	142.4	0.056	2.46	2.24
114.0	0.044	1.00	0.94	134.0	0.051	1.60	1.45
104.0	0.038	0.38	0.34	125.0	0.044	0.82	0.71

Test No. 125: Total conductor length, 29,050 cm.
Spacing, 91.4 cm.
No. 4 H. D. copper wire, diam. 0.518 cm.
Temperature, wet, 5.0 deg. C.
dry, 4.6 deg. C.
Barometer, 75.9 cm.
Cloudy, slight breeze.

Test No. 126: Total conductor length, 29,050 cm.
Spacing, 183 cm.
No. 4 H. D. copper wire, diam. 0.518 cm.
Temperature, wet, 5.0 deg. C.
dry, 4.5 deg. C.
Barometer, 75.9 cm.
Cloudy, slight breeze.

Test No. 137, Line *B*

Kv. bet. lines	Amp.	Kw.	Kw. line loss, *p*
80.0	....	0.05	0.02
90.5	0.025	0.11	0.06
100.5	0.029	0.35	0.26
110.7	0.037	0.95	0.78
121.0	0.044	1.42	1.19
131.0	0.051	2.11	1.76
141.5	0.056	2.70	2.26
150.8	0.064	3.24	2.72
161.0	0.072	4.05	3.40
172.0	0.078	4.80	4.05
183.0	0.084	5.60	4.73
196.0	0.093	6.60	5.59
205.0	0.102	7.60	6.44
202.0	0.098	7.30	6.19
186.0	0.087	6.00	5.08
165.0	0.075	4.30	3.63
145.0	0.061	2.86	2.40
124.0	0.047	1.75	1.46
103.0	0.032	0.60	0.47

Total conductor length, 29,050 cm.
Spacing, 366 cm.
No. 8 new H. D. copper wire, diam. 0.328 cm.
Temperature, wet, 1.5 deg. C.
dry, 1.5 deg. C.
Barometer, 76.6 cm.
Bright sun, slight breeze.

Test No. 138, Line *B*

Kv. bet. lines	Amp.	Kw.	Kw. line loss, *p*
79.2	....	0.06	0.03
91.2	0.025	0.12	0.07
99.9	0.027	0.26	0.18
111.4	0.036	0.08	0.65
120.8	0.039	1.22	1.06
121.5	0.049	1.90	1.59
141.0	0.055	2.30	2.20
149.0	0.059	2.80	2.34
161.0	0.066	3.40	2.85
171.4	0.074	4.20	3.53
181.4	0.079	4.80	4.05
192.0	0.085	5.60	4.73
202.2	0.092	6.56	5.67
214.4	0.11	7.50	6.35
197.0	0.089	6.10	5.15
174.0	0.076	4.40	3.71
153.2	0.063	3.00	2.51
134.4	0.051	2.00	1.67

Total conductor length, 29,050 cm.
Spacing, 488 cm.
No. 8 new H. D. copper wire, diam. 0.328 cm.
Temperature, wet, − 1.5 deg. C.
dry, + 1.5 deg. C.
Barometer, 75.5 cm.
Bright sun, slight breeze.

Test No. 92, Line *B*

Kv. bet. lines	Amp.	Kw.	Kw. line loss, *p*
27.5	0.008	....	....
34.5	0.009	....	....
39.5	0.011	....	....
44.5	0.013	0.02	0.01
51.0	0.015	0.05	0.04
56.5	0.017	0.10	0.09
61.5	0.019	0.22	0.20
66.5	0.023	0.37	0.31
71.0	0.024	0.49	0.40
76.0	0.028	0.60	0.49
83.0	0.032	0.81	0.66
90.5	0.037	1.07	0.88
101.0	0.041	1.43	1.17
110.5	0.050	1.80	1.46
120.5	0.055	2.30	1.88
131.5	0.060	2.80	2.28
144.5	0.068	3.50	2.84
158.0	0.081	4.40	3.57
170.0	0.086	5.30	4.33
181.0	0.094	6.18	5.15
190.0	0.099	6.70	5.43
204.0	0.110	8.00	6.50
215.0	0.117	9.00	7.30
222.0	0.123	9.64	7.84

Total conductor length, 29,050 cm.
Spacing, 410 cm.
0.066-in. galv. steel wire, diam. 168 cm.
Temperature, wet, 0.5 deg. C.
dry, 2.0 deg. C.
Barometer, 75.0
Cloudy, no wind.

Test No. 95, Line *B*

Kv. bet. lines	Amp.	Kw.	Kw. line loss, *p*
222.0	0.115	8.80	7.00
199.8	0.104	6.80	5.38
181.0	0.089	5.36	4.24
158.0	0.076	3.80	2.98
140.0	0.064	2.84	2.24
120.0	0.053	1.92	1.44
102.0	....	1.21	0.96
91.5	0.034	0.93	0.75
79.5	0.028	0.63	0.51
68.7	0.021	0.63	0.52
60.0	0.017	0.18	0.16
50.0	0.014	....	....

Total conductor length, 29,050 cm.
Spacing, 550 cm.
0.066 in. galv. steel wire, diam. 168 cm.
Temperature, wet, 1.0 deg. C.
dry, 3.0 deg. C.
Barometer, 75.0 cm.
Cloudy, no wind.

TRANSFORMER LOSS

Kv.	Amperes	Kw.	Kv.	Amperes	Kw.
101.5	0.008		71.5	0.005	0.02
131.5	0.010	0.15	82.0	0.006	0.03
147.3	0.011	0.25	97.0	0.007	0.05
			112.0	0.008	0.06
163.8	0.013	0.42	132.8	0.009	0.09
181.5	0.014	0.54			
201.8	0.016	0.69	149.0	0.010	0.12
			178.4	0.013	0.18
	30° C.		201.0	0.014	0.22
			223.0	0.016	0.30
				3° C.	

INDEX

A

B

C

D

E

17

S

CPSIA information can be obtained at www.ICGtesting.com
Printed in the USA
LVOW121001101012

302239LV00004B/81/P

9 781603 861120